Disclaimer

The publisher of this book is by no way associated with the National Institute of Standards and Technology (NIST). The NIST did not publish this book. It was published by 50 page publications under the public domain license.

50 Page Publications.

Book Title: Exposure and Fire Hazard Assessment of Nanoparticles in Fire Safe Consumer Products: Interagency Agreement Final Report

Book Author: Rick D. Davis; Yeon Seok Kim; Richard H. Harris; Marc R. Nyden; Nasir M. Uddin; Treye Thomas

Book Abstract: An innovative technology was evaluated to generate fire and health safe soft furnishings. Nanoparticle-based thin coatings on a polyurethane foam and nonwoven barrier fabric were applied using Layer-by-layer (Lbl) assembly. This is the first report of using Lbl on a complex three dimensional substrate, to improve the fire resistance of foam and barrier fabrics, and with sodium montmorillonite clay (MMT), carbon nanofibers (CNF) and multi-walled carbon nanotubes (MWCNT). The Lbl process was tailored for each nanoparticle in order to fabricate durable coatings that completely covered the entire substrate surface. The CNF and MWCNT coatings on foam were thinner and contained fewer nanoparticles, but resulted in the greatest reduction in peak heat release rate (flammability). The reduction in foam flammability due to the nanoparticle \Lbl coatings is as high as 1138% greater than 17 other commercial fire retardants commonly used in foam. This technology has strong commercial viability for foam due to easy and flexibility of the Lbl process and the significant reduction in foam flammability caused by the coatings. However, Lbl does not work for nonwoven barrier fabrics as the structure was unable to remain intact during the fabrication process. In order to enable other agencies to access the potential health risk of using this nanoparticle-based technology for reducing the flammability of soft consumer products, this project developed the methodology to promote, collect, and quantify nanoparticles released from these substrates. In general, the release of nanoparticles was an order of magnitude higher from simulated chewing than simulated wear and tear, highest from the barrier fabric, and lowest for MMT. The release was between 0.50 mass fraction % to 0.0003 mass fraction % of the total nanoparticle loading on the substrate.

Citation: NIST Interagency/Internal Report (NISTIR) - 7805

Keyword: nanoparticle; flammability; foam; mattress; upholstered furniture; service-life; health; risk exposure

NIST Internal Report 7805

Interagency Agreement Report: Exposure and Fire Hazard Assessment of Nanoparticles in Fire Safe Consumer Products

Rick D. Davis
Yeon Seok Kim
Richard H. Harris
Marc Nyden
Nasir Uddin
Treye Thomas

August 2011

NIST National Institute of Standards and Technology • U.S. Department of Commerce

NIST Internal Report 7805

Interagency Agreement Report: Exposure and Fire Hazard Assessment of Nanoparticles in Fire Safe Consumer Products

Rick D. Davis
Yeon Seok Kim
Richard H. Harris
Marc Nyden
Nasir Uddin
Fire Research Division
Engineering Laboratory
National Institute of Standards and Technology

Treye Thomas
Directorate of Health Services
Consumer Products Safety Commission

August 2011

U.S. Department of Commerce
Gary Locke, Secretary

National Institute of Standards and Technology
Patrick D. Gallagher, Director

Certain commercial entities, equipment, or materials may be identified in this document in order to describe an experimental procedure or concept adequately. Such identification is not intended to imply recommendation or endorsement by the National Institute of Standards and Technology, nor is it intended to imply that the entities, materials, or equipment are necessarily the best available for the purpose.

National Institute of Standards and Technology and Internal Report
Natl. Inst. Stand. Techn. Internal Report 7805, 57 pages (August 2011)
CODEN: NTNUE2

Abstract

An innovative technology was evaluated as a potential approach to generate flammability, environmental, health, and safety compliant soft furnishings. Nanoparticle-based nanometer thin film coatings on standard polyurethane foam and non-woven fire blocking barrier fabrics were applied using a waterborne approach called Layer-by-layer (LbL) assembly. This is the first reportof LbL on a complex three dimensional substrate to improve the fire resistance of foam and barrier fabrics using sodium montmorillonite clay (MMT), carbon nanofibers (CNF), and multi-walled carbon nanotubes (MWCNT). The LbL process and recipe was tailored for each nanoparticle type (CNF, MWCNT, and MMT) in order to fabricate durable coatings that completely covered the entire internal and external surfaces of the substrate. The CNF and MWCNT coatings on foam were thinner and contained fewer nanoparticles, but resulted in the greatest reduction in peak heat release rate (a measure of material flammability). The reduction in foam peak heat release due to the nano-fire retardant\LbL coatings is as high as 1100 % greater than 17 other commercial fire retardants commonly used to reduce foam flammability. This technology has strong commercial viability for foam due to ease and flexibility of the LbL process and the significant reduction in foam flammability caused by the coatings. However, does not appear to be a valid approach for barrier fabrics as the non-woven structure disassembled during the fabrication process.

In order to enable other agencies to assess the environmental, health, and safety risk of using this nanoparticle-based technology for reducing the flammability of soft consumer products, this project developed the methodology to promote, collect, and quantify nanoparticles released from various substrates when exposed to various conditions. In general, the release of nanoparticles was an order of magnitude higher from simulated chewing than simulated wear and tear, highest from the barrier fabric, and lowest for MMT. The release was between 0.50 mass fraction % to 0.0003 mass fraction % of the total nanoparticle loading on the substrate. The measurement methodology developed enables the quantification of nanoparticles with greater accuracy and at an order of magnitude lower concentration than current measurement methods.

Keywords

Nanoparticle; flammability; foam; mattress; upholstered furniture; service-life; fire retardants; Layer-by-layer assembly; cone calorimetry; heat release rate; carbon nanofiber; carbon nanotubes; clay, barrier fabric; Consumer Product Safety Commission

Acknowledgements

Appreciation is extended to the Consumer Product Safety Commission for financial and technical support for this project. Interagency Agreement Number: CPSC-I-08-0008

Page Intentionally Left Blank

Contents

Abstract	Error! Bookmark not defined.
Acknowledgements	Error! Bookmark not defined.
List of Tables	ix
List of Figures	x
List of Acronyms	Error! Bookmark not defined.
1. Introduction	15
2. Experimental	17
2.1. Materials	17
2.1.2. CNF-based bilayered nanocoatings	19
2.1.3. MWCNT-based trilayered nanocoatings	20
2.1.4. MMT-based trilayered nanocoatings	21
2.2. Fire performance testing	22
2.3. Simulated aging	23
2.2. Characterization	24
3. Results and Discussion	26
3.1. Characterization of nanocoatings	28
3.1.1. CNF-based nanocoatings	29
3.1.2. MWCNT-based nanocoatings	32
3.1.2. MMT-based nanocoatings	36
3.2. Fire performance	39
3.2.1. CNF-based nanocoatings	39
3.2.2. MWCNT-based nanocoatings	40
3.2.3. MMT-based nanocoatings	41
4.4. Comparison to other flame retarding technologies	44
Flame Retardant Location	46
% Reduction relative to PUF	46
In coating on foam	46
Embedded in foam	46
3.5. Nanoparticle release from stressing	46
3.5.1. Measurements methodology	46
3.5.2. Stress induced nanoparticle release	50

4. Conclusion 52
5. Future Research 54
6. References 54

List of Tables

Table 1. Provided are the average physical characteristics of nanoparticle coated substrates organized by substrate type and highest to lowest in nanoparticle content. BFs coating thickness is assumed to be similar to that measured on PUF. MMT coatings are sufficiently different on PUF and BF that the coating thickness on PUF is believed to not be a good estimate of the BF coating thickness. Values reported with 2σ uncertainty. 28

Table 2. Cone Calorimetry data of the washed uncoated and coated PUF organized from highest to lowest in PHRR (**PHRR in bold**). Values reported with 2σ uncertainty................................ 42

Table 3. Cone Calorimetry data of the uncoated and coated BF organized from highest to lowest in PHRR. Values reported with 2σ uncertainty. 43

Table 4. Reduction in Cone data (relative to pure PUF) caused by FR in a LbL coating on and embedded in foam. Adjusting for different FR loadings, the LbL approach can result in the largest reduction in PHRR. No uncertainty was reported for the literature values. Experimental values reported with 2σ uncertainty... 46

Table 5. Simulated wear and tear nanoparticles released relative to the total nanoparticle content on the specimen (organized from highest to lowest in nanoparticle release). Values reported with 2σ uncertainty. 52

Table 6. Simulated chewing nanoparticles released relative to the total nanoparticle content on the specimen (organized from highest to lowest in nanoparticle release). Values reported with 2σ uncertainty... 52

List of Figures

Figure 1. The CNF/polymer coating process was an alternating submersion in a cationic (CNF/PEI) and an anionic (PAA) solution with washing (rinse and wring) between each solution. After creating 4 BLs (BL is CNF:PEI/PAA), the specimen was dried in a convection oven for. 19

Figure 2. The MWCNT/polymer coating process was an alternating submersion in an anionic (PAA), MWCNT cationic (MWCNT-PEI) and polymer cationic (PEI only) solutions with washing (rinse and wring) between each solution. After creating four TLs (TL is PAA\MWCNT-PEI\PEI), the specimen was dried in a convection oven for 12 h at 70 °C ± 1 °C to remove excess water. 21

Figure 3. The MMT/polymer coating process was an alternating submersion in a polymer cationic (PEI), clay anionic (MMT), and polymeric anionic (PAA) solutions with washing (rinse and wring) between each solution. After creating five TLs (TL is PEI\MMT\PAA), the specimen was dried in a convection oven for 12 h at 70 °C ± 1 °C to remove excess water. 22

Figure 4. Simulated saliva extractions were performed using bottles containing a simulated saliva and a substrate that were rotated at 6.3 rad/s (60 rev/min) for 30 min. 24

Figure 5. Simulated normal wear and tear stressing was performed by pounding substrates at 1 cycle/s for 100,000 cycles at 20682 Pa ± 69 Pa of pressure. 24

Figure 6. SEM images of as-received BF (a) 20,000x, (b) 50,000x, (c) 100,000x, and (d) 200,000x. The 27 µm ± 3 µm diamter strand is a bundle of several fibers that "welded"during manufacturing. 26

Figure 7. SEM images of as-received PUF at (a) 1,000x, (b) 2,000x, (c) 5,000x, (d) 10,000x, (e) 20,000x, (f) 50,000x and (g) 100,000x and washed PUF at (h) 5,000x. The PUF surface was smooth and featureless after debris (dust, etc.) was removed by washing (h). 27

Figure 8. SEM images of the inside section of a CNF/PUF at (a) 1,000x, (b) 5,000x, (c) 10,000x, and (d) 20,000x, of a thicker island at (e) 50,000x (f) 100,000x, and (g) 200,000x, and of an aggregate at (h) 200,000x. 30

Figure 9. SEM images of a fractured edge of CNF/PUF at (a) 20,000x, (b) 100,000x, (c) 200,000x, and (d) 500,000x. The CNF coating is 359 nm ± 36 nm. 31

Figure 10. SEM images of a delaminated CNF/PUF at (a) 10,000x, (b) 50,000x, (c) 100,000x, and (d) 200,000x. The CNFs below the surface are welded together with polymer. The root cause of delimination may be the freeze fracture process or poor adhesion to the PUF due to the high CNF concentration. 31

Figure 11. SEM images of the inside section of a CNF/BF at (a) 20,000x, (b) 50,000x, (c) 100,000x, and (d) 200,000x and (e) 200,000x. The CNF coatings on the PUFs and BFs appear to be similar. 32

Figure 12. SEM images of the inside section of a MWCNT/PUF at (a) 1,000x, (b) 5,000x, (c) 10,000x, (d) 20,000x, (e) 50,000x (f) 100,000x, (g) 200,000x, and (h) 500,000x. The MWCNTs were well dispersed and distributed throughout the polymer coating. The coating was smooth and featureless except for a few small larger aggregates and a few 10 nm ± 5 nm wide cracks.. 34

Figure 13. SEM images of a fractured edge of MWCNT/PUF at (a) 50,000x, (b) 100,000x, (c) 200,000x, and (d) 500,000x. The MWCNT coating thickness is 440 nm ± 47 nm. 35

Figure 14. SEM images of the inside section of a MWCNT/BF (a) 20,000x, (b) 50,000x, (c) 100,000x, and (d) 200,000x. The MWCNT coatings appear similar on PUF and BF. 35

Figure 15. SEM images of the inside section of a MMT/PUF at (a) 1,000x, (b) 5,000x, (c) 10,000x, (d) 20,000x, (e) 50,000x (f) 100,000x, and (g) 200,000x. 37

Figure 16. SEM images of a fractured edge of MMT/PUFat (a) 10,000x, (b) 20,000x, (c) 50,000x, and (d) 100,000x. The MMT coating was 1000 nm ± 450 nm. 38

Figure 17. SEM images of a MMT/BF (a) 10,000x, (b) 20,000x, (c) 50,000x, (d) 100,000x and (e) 200,000x. The MMT coatings are on PUF and BF. 38

Figure 18. HRR curves indicate the MWCNT and CNF coatings significantly reduce PHRR, THR, and total burn time, but MMT coatings increase flammability, as compared to pure PUF. The 2σ uncertainty is ± 5% in HRR and ± 2 s in time. 43

Figure 19. HRR curves of the uncoated and coated BF. All nanoparticle coatings deteriorate the fire performance of the BF. The 2σ uncertainty is ± 5% in HRR and ± 2 s in time. 44

Figure 20. UV-VIS absorbance plot of a CNF/DI/SDS suspension prepared without sodium chloride, with sodium chloride, and the sodium chloride removed by dialysis. Dialysis was critical to obtaining accurate and repeatable absorbance values at the MWCNT and CNF peak maximum (267 nm) for the simulated chewing suspensions. 47

Figure 21. A Beer's Law curve was constructed from the UV-VIS absorbance value of several CNF or MWCNT calibration standards (CNF provided here). CNF and MWCNT concentration was quantified based on the measured UV-VIS absorbance value of a stressed suspension and a Beer's Law curve. The 2σ uncertainty is ± 5 % of the CNF or MWCNT mass fraction value... 48

Figure 22. ICP-OES intensity as a function of element type and concentration using a diluted nitric acid/DI solution containing 0.0010 mass fraction % ± 0.0005 mass fraction % MMT and five different concentrations of each element. The concentration at an intensity of 0 cps is the amount of each element in MMT. The 2σ uncertainty is 5 % of the value (error bars are smaller than the data markers). 49

Figure 23. The amount of each element in MMT in a diluted nitric acid/DI solution containing 0.0010 mass fraction % ± 0.0005 mass fraction % MMT in diluted. The error bars represent a 2σ standard uncertainty in the concentration. 49

Figure 24. Beer's Law calibration curve of ICP-OES intensity as a function of MMT concentration. The 2σ uncertainty is 5 % of the value (error bars are smaller than the data markers). 50

List of Acronyms

ANPR	Announced notice of proposed rulemaking
ASTM	American Society for Testing and Materials
BF	Barrier fabric
BL	Bilayer
CNF	Carbon nanofibers
CNF/BF	Carbon nanofiber coated barrier fabric
CNF/PUF	Carbon nanofiber coated standard polyurethane foam
CPSC	Consumer Product Safety Commission
DI	Deionized (water)
DMF	*N,N*-dimethylformamide
ICP-OES	Inductively Coupled Plasma – Optical Emission Spectrometer
LbL	Layer-by-layer assembly/coating
MMT	Montmorillonite clay
MMT/BF	Montmorillonite clay coated barrier fabric
MMT/PUF	Montmorillonite clay coated standard polyurethane foam
MWCNT	Multi-walled carbon nanotubes
MWCNT/BF	Multi-walled carbon nanotubes coated barrier fabric
MWCNT/PUF	Multi-walled carbon nanotubes coated standard polyurethane foam
NIST	National Institute of Standards and Technology
PAA	Poly(acrylic acid)
PEI	Branched Polyethyleneimine
PHRR	Peak heat release rate
PUF	Standard polyurethane foam
SEM	Scanning Electron Microscopy
TGA	Thermal gravimetric analysis
THR	Total heat released
TL	Trilayer

Page Intentionally Left Blank

1. Introduction

The estimated annual total societal cost of fire to the United States economy is $360 billion [1] with fires in structures, such as single and multi-family dwellings, and fixed mobile homes, accounting for an estimated $90 billion of this total cost [2]. There are a reported 135,000 residential home fires annually, 5 % of which are accounted for by soft furnishings (mattresses, bedclothes, upholstered furniture) as the first item ignited. However, first item ignited soft furnishings are annually estimated to account for a disproportionately high amount of fire losses (33 % of the civilian fatalities, 18 % of civilian injuries, and 11 % of the property losses) [3,4]. Even though no amount of money can adequately represent personal injury and deaths, the estimated annual societal cost of fire associated with soft furnishings as the first item ignited is estimated at $5 billion based on $5 million per fatality and $230,000 per injury as used by Hall [1] in calculating the total societal cost of fire.

The Consumer Products Safety Commission (CPSC) is responsible for existing and proposed United States flammability regulations of soft furnishings [5,6,7]. These regulations, as well as the introduction of Reduced Ignition Propensity cigarettes [8], are expected to significantly reduce the $5 billion annual cost of soft furnishing fires. In order to comply with the 2007 open flame mattress regulation (CPSC 16 CFR 1633 [6]), manufacturers inserted fire blocking barrier fabrics around the soft polyurethane foam core. CPSC research (unpublished) suggests that upholstered furniture manufacturer's will likely also require fire blocking barrier fabrics in order to comply with the proposed open flame/smoldering ignition regulation for upholstered furniture (CPSC ANPR 16 CFR 1634 [7]). Similar to manufacturer of other fire safe products, the technical and engineering options to comply with national and/or international fire performance regulations are quickly diminishing because of mandated sustainability regulations for consumer products (e.g., REACH [9] and EcoLabel [10]).

Layer-by-Layer (LbL) assembly has been extensively studied for the past 20 year as a methodology to create multifunctional films generally less than 1μm thick [11,12,13]. The thin film coatings were commonly fabricated by alternate deposition of a positively charged layer and negatively charge layer (called a bilayer, BL). By taking advantage of electrostatic, H-bonding [14], covalent bonds [15], and/or donor/acceptor interactions, these bilayers are continuously assembled on the surface of flat substrates. The LbL process is quite flexible and robust, which allows it to be tuned for specific coating characteristics and for coating a range of substrate types. For example, altering the concentration, pH, and/or temperature of the LbL solutions can result in a 1 nm rather than 100 nm thick BL [16,17]. LbL thin films have been used in an extensive breadth of applications, such as oxygen barriers [18] and sensors [19], and have useful properties, such as antimicrobial [20] and antireflection [21]. A more recent application, which is directly aligned with the research presented in this manuscript, was LbL montmorillonite (MMT) coatings (sodium exchanged MMT) of cotton fabric to improve the fire performance characteristics of this textile [13]. MMT has been extensively studied in LbL thin films [18,22,23] and, when used as an additive filler, has been shown to simultaneously improve the mechanical and fire performance attributes of polymers [24,25,26]. The uniqueness this approach is the concept of improving fire performance by using LbL assembly and creating LbL MMT coatings on cotton fabric (MMT/cotton) [13]. The results are exciting in that Li achieved complete and uniform high quality MMT based coatings on cotton. In addition, the MMT coatings resulted in a significant retention of fabric like char after conducting vertical burn tests and there was no or less ember afterglow when the flame was removed. These results suggest the coating may better prevent thermal and flame penetration from reaching and igniting the foam (PUF), and therefore, the MMT/cotton may reduce the risk of fire spread in residential homes if used in soft furnishings.

Natural and synthetic clays are one of the most widely studied additives to improve polymer fire performance through the formation of a nanocomposite [27]. The ease of functionalizing the clay with a range of different organic treatments is the main reason for clay's versatility and popularity as a polymer additive. Smectite clays are perhaps the most commonly used because they are water dispersible making them ideal for organic modification. In contrast to other fire retardants, clay at a minimum can often improve fire performance while maintaining other polymer physical and mechanical attributes of the polymer. This is attributed to both the significantly lower loading levels of clay (compared to other flame retardants) to obtain the same performance and the interfacial interactions between the clay and polymer. Synergistically combining flame retardants with clay is not necessary intuitive. For example, a 2 mass fraction % MMT added to polystyrene with (10, 15, or 20) mass fraction % ammonium polyphosphate and pentaerythriol (APP/PER) resulted in a reduce flame spread rate as compared to a similar loading level of just APP/PER. In contrast, all other combinations ((4, 6, 8, and 10) mass fraction % MMT with (5, 10, 15, 20, 25, or 30) mass fraction % APP/PER) resulted in an increased flame spread rate (as high as 20% faster) as compared to a similar loading level of just APP/PER [28].

Carbon nanofibers (CNFs) are cylindrical nanostructures constructed of stacked graphitic cones or cups. Compared to carbon nanotubes (CNT), CNFs can be at least an order of magnitude larger with a diameter and length in the range of 5 nm to 300 nm and 0.1 μm to 1000 μm, respectively. Due to the intrinsic electrical, thermal, and mechanical properties of CNFs, the thermal and electrical conductivity, tensile and compressive strength, ablation resistance, damping properties, and flammability of polymers [29] have been significantly altered with incorporation of CNF [30].

Recently, a reduction in PUF flammability was reported by the incorporation of CNFs directly into the PUF [29]. At a 4 mass fraction % CNF loading, the CNFs formed a network structure that reduced the peak heat release rate (PHRR) by 35% and prevented melt dripping. The approach of incorporating CNFs into the PUF has a few potential drawbacks. For example, commercialization may be difficult as the foam manufacturing process is extremely sensitive to small changes in recipe, especially the presence of solid particles. Another potential drawback is based on the mechanism by which nanoparticles are believed to reduce polymer flammability [31]. It is has been proposed the reduction in flammability primarily results from the formation of a char at the surface that thermally protects the polymer and prevents volatilization of polymer degradation products. Perhaps placing the nanoparticles at the surface could accelerate char formation.

Compared to CNFs, CNT are significantly smaller in dimension. Single-walled carbon nanotubes (SWCNT) and multi-walled carbon nanotubes (MWCNT) have received a lot of attention in the science community because their inherent characteristics, such as small size and high aspect ratio (diameter of 1 nm to 100 nm and length of 1 μm to 100 μm for SWCNT and MWCNT, respectively) [32,33], high modulus (approximately 1 TPa) [34], high intrinsic electrical conductivity ($\sigma > 10^4$ S/cm) [35], and high thermal conductivity ($k > 1000$ W/m·K) [36], are expected to impart electrical conductivity, mechanical strength, and thermal conductivity to a polymer when the CNTs are incorporated into the polymer. However, CNTs continue to be underutilized partially due to the difficulties in generating stabilized CNT suspensions. To improve stability researchers have used non-covalent stabilizing agents (e.g., surfactants [37,38,39], water-soluble polymers [40,41,42], and inorganic nanoparticles [43,44]) and chemically modified CNTs. For electrical conductivity, non-covalent stabilization of CNTs is preferred over covalent functionalization because chemical modification has been shown to reduce conductivity [45,46,47]. On the other hand, covalent functionalization shows better solubility due to the strong interfacial interaction between the nanotubes and polymer matrix via direct chemical bonding.

The research presented in this report is unique in that it is the first published report of fabricating CNF, MWCNT, and MMT based thin films/coatings on foam and barrier fabrics with the intent to improve their fire performance, and measuring the release of nanoparticles from these coated substrates under conditions that mimic the end-use conditions of soft furnishings. The LbL approach, we believe, is ideal for reducing the flammability of the substrates, and by extension soft furnishing, as it will more quickly form the char-like armor needed to reduce the flammability of soft furnishings because the high concentration of nanoparticles are already at the surface (of the foam and barrier fabric) rather than randomly mixed throughout the polymer. This approach is also more commercially viable than impregnating the substrates with nanoparticles as it is a post-manufacturing process that uses equipment commonly used in the manufacturing of these substrates. Industry has expressed interest in using this technique.

2. Experimental [48,49,50]

Unless otherwise indicated all values are reported with 2σ uncertainty.

2.1. Materials

All materials were used as-received from the supplier unless otherwise indicated. Branched polyethylenimine (PEI, branched, Molecular Mass = 25,000 g/mol) and poly(acrylic acid) (PAA, Molecular Mass = 100,000 g/mol) were obtained from Sigma-Aldrich (Milwaukee, WI). These polymers were selected primarily because their behavior in LbL assembly is well documented and understood. Baytubes C150HP multi-walled carbon nanotubes (MWCNT, average diameter was 14 nm, length was 1 μm to 10 μm) were obtained from Bayer MaterialScience AG (Pittsburgh, Pennsylvania). PR-24-XT-PS carbon nanofibers (CNF, average diameter = 100 nm, length was 30 μm to 100 μm) were obtained from Pyrograf Products Incorporated. Sodium montmorillonite clay (MMT, trade name is Cloisite Na$^+$) was obtained from Southern Clay Products Inc. (Gonzales, TX). The standard (untreated) polyurethane foam [51] coated in this study was stored as-received from the supplier (cardboard box with no packaging material at 25 °C ± 2 °C). On the day of coating, nine substrates (length/width//height of (10.2 cm / 10.2 cm / 5.1 cm) ± 0.1 cm) were cut from a single substrate (length/width//height of (30.6 cm / 30.6 cm / 5.1 cm) ± 0.1 cm). These smaller substrates (0.6 mass fraction % ± 0.1 mass fraction %) were rinsed and wringed out (discussed below in the coating process) to remove debris and other extractables. After drying, the post-extraction mass of these substrates was 12.7 g ± 0.3 g. The barrier fabric (BF) coated in this study was a poly(melamine-co-formaldehyde) based non-woven (density of 222 g/m^2). The BFs were not washed prior to coating because there was no visible evidence of dust or debris, and the integrity of the BF deteriorates after 10 washings.

The polyelectrolyte (0.1 mass fraction % ± 0.03 mass fraction %) and DI (< 0.5 μS) water solutions were prepared as follows. A 2 L glass container was charged with DI water (1300 mL) and PEI (0.1 mass fraction % ± 0.03 mass fraction %, 1.3 g ± 0.4 g). This PEI cationic stock solution was slowly agitated for 6 h at room temperature before using. The preparation of the PAA anionic stock solution was similar to the PEI cationic solution, except PAA (0.1 mass fraction % ± 0.03 mass fraction %, 1.3 g ± 0.4 g) was used instead of PEI. The pH value was 10 and 3 for the PEI and PAA solutions, respectively.

The CNF/PEI suspension in DI water was prepared by charging a plastic bottle (250 mL) with the PEI cationic stock solution (150 mL ± 1 mL) then adding CNF powder (0.050 mass fraction % ± 0.003 mass fraction % relative to total PEI stock solution (600 mL), 0.30 g ± 0.02 g). The suspension was sonicated at 40 W using a Sonics VCX130 sonicator with a 13 mm probe for 1 h with the temperature never exceeding 70 °C ± 1 °C. The sonicated suspension was diluted with more PEI stock solution (450 mL) and was manually agitated for 3 min ± 1 min. The CNF/PEI suspension was immediately used for coating.

The MWCNTs were first functionalized with PEI to facilitate dispersion and distribution in DI water and to improve retention of the MWCNTs in the coating. Amination of MWCNTs was prepared according to the procedure by Liao et al [52]. A plastic vial (500 mL) was charged with *N,N*-dimethylformamide (200 mL ± 1 mL, DMF), PEI (5.0 mass fraction % ± 0.1 mass fraction relative to DMF, 10 g ± 0.1 g), and MWCNT (20.0 mass fraction % ± 0.1 mass fraction % relative to PEI, 2.0 g ± 0.1 g), The mixture was sonicated at 40 watts using a Sonics VCX130 sonicator with a 13 mm probe for 1 h then agitated with a stir bar for 2 d at 50 °C ± 2 °C. The functionalized MWCNT product (MWCNT-PEI) was isolated from the suspension by filtering through a 0.20 μm nylon membrane and washing four times with alternating methanol and water washes to remove excess PEI and DMF. The MWCNT-PEIs were dried in a dessicator with anhydrous calcium sulfate for at least 3 d prior to use. Once dried, MWCNT-PEIs were grounded using mortar and pestel.

The MWCNT-PEI in DI water suspension was prepared by charging a plastic bottle (250 mL) with DI water (150 mL ± 1 mL) and MWCNT-PEI (0.10 mass fraction % ± 0.03 mass fraction % relative to total DI water, 0.60 g ± 0.02 g). The suspension was sonicated at 40 W using a Sonics VCX130 sonicator with a 13 mm probe for 1 h with the temperature never exceeding 70 °C ± 1 °C then was diluted with more DI water (450 mL) and shaken by hand for 3 min ± 1 min. The MWCNT-PEI suspension was used immediately for coating.

The MMT in DI water suspension was prepared by charging a 2 L glass container with DI water (1300 mL) then adding MMT (0.2 mass fraction % ± 0.03 mass fraction %, 2.6 g ± 0.8 g). This MMT suspension was stirred for 12 h using a magnetic stirrer before coating. The MMT suspension was used immediately for coating.

2.1.2. CNF-based bilayered nanocoatings

The four BL CNF coatings on PUF and BF took approximately 30 min per specimen (14 min for first BL and 15 min for the remaining three BLs). In general, the fabrication process was alternately depositing cationic (CNF and PEI) and anionic (PAA) layers on the surface of the substrate and removing unbound material (polymer and CNF) by rinsing with DI water and wringing out the excess water several times (

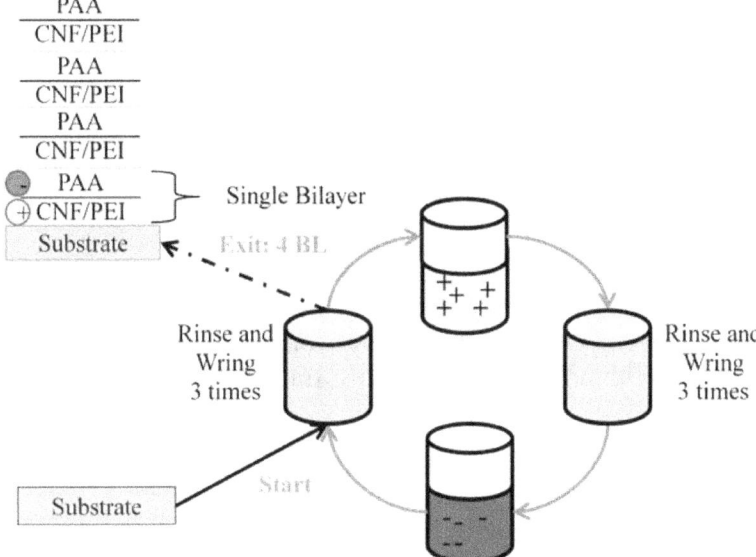

Figure 1). The process of removing excess water using a convection oven and dessicator extended over a period of 3 d.

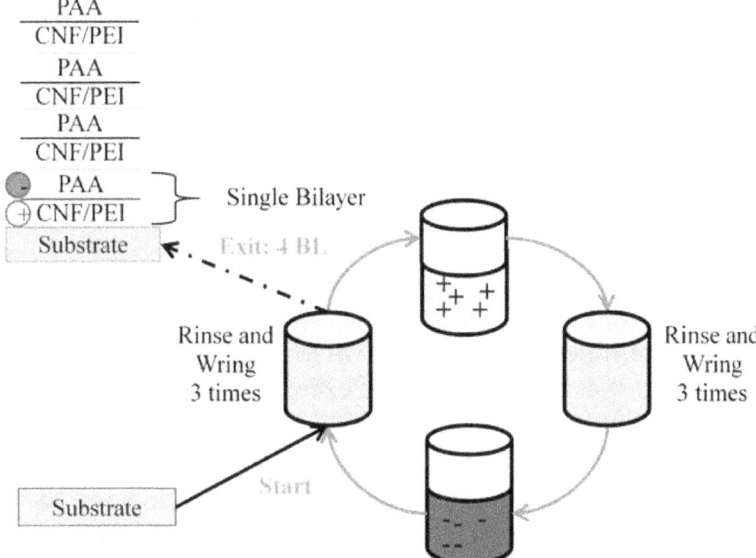

Figure 1. The CNF/polymer coating process was an alternating submersion in a cationic (CNF/PEI) and an anionic (PAA) solution with washing (rinse and wring) between each solution. After creating 4 BLs (BL is CNF:PEI/PAA), the specimen was dried in a convection oven for 12 h at 70 °C ± 1 °C to remove excess water.

More specifically, a plastic container (2 L) was charged with the CNF:PEI cationic suspension (600 mL ± 10 mL), a similar container was charged with the PAA anionic solution (600 mL ± 10 mL), and six more containers were charged with deionized water (600 mL ± 10 mL each). The substrate was submersed into the CNF:PEI cationic suspension and after squeezing and releasing the substrate four times in the CNF:PEI suspensions, the substrate was soaked in the suspension for an additional 5 min. The substrate was removed and the excess solution was squeezed back into the cationic dipping container. To remove unbound PEI and/or CNF, the substrate was thoroughly rinsed in three separate containers. Since most of the cationic materials were typically removed in the first rinsing container, the rinsing water in this container was replaced with fresh deionized water after each washing cycle. Excess water was removed by passing the substrate twice through a manually powered wringer. The PAA anionic layer was then deposited and the unbound PAA was removed using the same procedure as described above, except the washing was performed using different rinsing containers. This deposition of the CNF:PEI layer followed by the PAA layer created a single BL (CNF:PEI/PAA). The procedure for depositing the next three BLs was similar to the first BL, except the substrate was only submersed in the coating solutions for 1 min rather than 5 min. After the four BLs were deposited, the specimen was dried in a convection oven (70 °C ± 1 °C, 12 h) and stored in a dessicator (at least 3 d) with anhydrous calcium sulfate before weighing and analyzing.

2.1.3. MWCNT-based trilayered nanocoatings

The four trilayer (TL) MWCNT coatings on PUF and BF took approximately 40 min per specimen (20 min for the first TL and 20 min for the remaining three TL). In general, the fabrication process was alternately depositing an anionic layer (PAA), a functionalized MWCNT only cationic layer (MWCNT-PEI), and a polymer only cationic layer (PEI) on the surface of the substrate and removing unbound material (MWCNT and polymer) by rinsing with DI water and wringing out the excess water several times (Figure 2). The process of removing excess water using a convection oven and dessicator occurred over a period of 3 d.

More specifically, three plastic containers (2 L) were charged with the coating solutions. One container (2 L) was charged with the PAA anionic solution (600 mL ± 10 mL), a second with the MWCNT-PEI cationic suspension (600 mL ± 10 mL), and a third with the PEI cationic solution (600 mL ± 10 mL). Three rinsing containers (2 L) per coating solution were charged with DI water (600 mL ± 10 mL, each). The substrate was submersed into the PAA anionic solution and after squeezing and releasing the substrate four times, the substrate was soaked in the PAA solution for an additional 5 min. The substrate was removed and the excess solution was squeezed back into the anionic dipping container. To remove unbound PAA, the substrate was thoroughly rinsed in three separate containers. Since most of the PAA was typically removed in the first rinsing container, the rinsing water in this container was replaced with fresh deionized water after each washing cycle. Excess water was removed by passing the substrate twice through Dyna-Jet BL-44 hand wringer (Dyna-Jet Products, Overland Park, KS).

The MWCNT-PEI cationic layer was then deposited onto the PAA/substrate and the unbound MWCNT-PEI was removed using the same procedure described above, except the washings were performed using different rinsing containers. The polymer only cationic layer (PEI only) was then deposited onto the MWCNT-PEI/PAA/substrate and washed using the same procedures described above, except using different rinsing containers. This deposition of a PAA layer, a MWCNT-PEI layer, and a PEI layered created a single TL (PEI/MWCNT-PEI/PAA/substrate). The procedure for depositing the next three TLs was the same as the first TL, except the substrate was only soaked in the coating solutions for 1 min rather than 5 min. After the coating was complete, the specimen was dried in a convection oven (70 °C ± 1 °C, 12 h) and then stored in a dessicator (at least 3 d) with anhydrous calcium sulfate before weighing and analyzing.

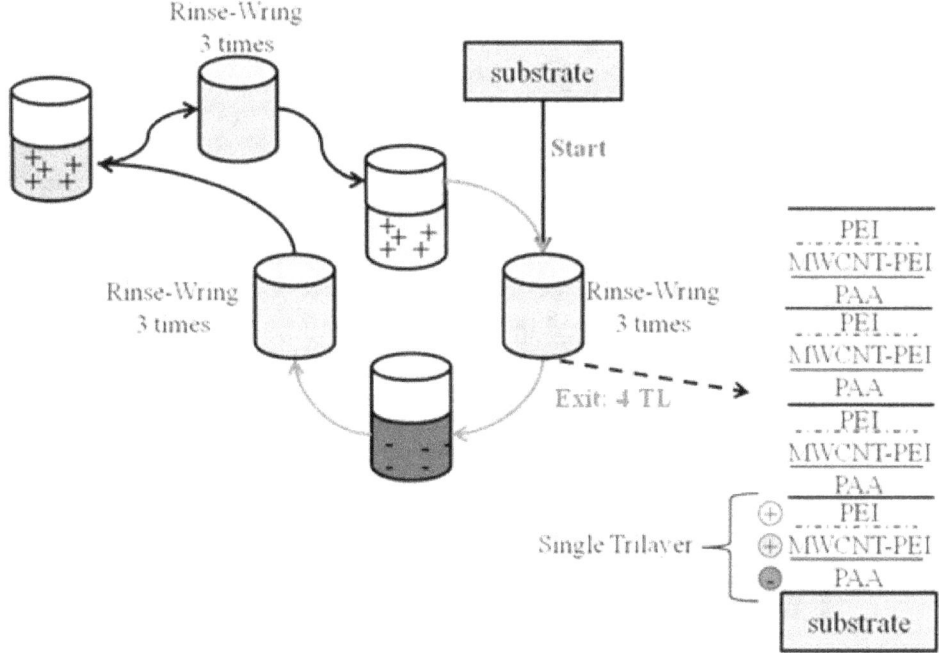

Figure 2. The MWCNT/polymer coating process was an alternating submersion in an anionic (PAA), MWCNT cationic (MWCNT-PEI) and polymer cationic (PEI only) solutions with washing (rinse and wring) between each solution. After creating four TLs (TL is PAA\MWCNT-PEI\PEI), the specimen was dried in a convection oven for 12 h at 70 °C ± 1 °C to remove excess water.

2.1.4. MMT-based trilayered nanocoatings

The eight TL MMT coatings on PUF and BF took approximately 50 min per specimen (20 min for the first TL and 30 min for the remaining seven TL). In general, the fabrication process was alternately depositing a cationic layer (PEI), an anionic clay layer (MMT), and an anionic layer (PAA) on the surface of the substrate and removing unbound material (MMT and polymers) by rinsing with DI water and wringing out the excess water several times (3). Similar to the CNF and MWCNT coating processes, the excess water was removed over 3 d using a convection oven and a dessicator.

More specifically, three plastic containers (2 L) were charged with the coating solutions. One container (2 L) was charged with the PEI cationic solution (600 mL ± 10 mL), a second with the MMT anionic suspension (600 mL ± 10 mL), and a third with the PAA anionic solution

(600 mL ± 10 mL) (3). Three rinsing containers (2 L) per coating solution were charged with DI water (600 mL ± 10 mL, each). A substrate was submersed into the PEI cationic soultion and after squeezing and releasing the substrate four times, the substrate soaked in the PEI solution for an additional 5 min. The substrate was removed and the excess solution was squeezed back into the anionic dipping container. To remove unbound PEI, the substrate was thoroughly rinsed in three separate containers. Since most of the PEI was typically removed in the first rinsing container, the rinsing water in this container was replaced with fresh deionized water after each washing cycle. Excess water was removed by passing the substrate twice through a Dyna-Jet BL-44 hand wringer (Dyna-Jet Products, Overland Park, KS).

The MMT anionic layer was then deposited onto the PEI/substrate and the unbound MMT was removed using the same procedure described above, except the washings were performed using different rinsing containers. The second anionic (PAA only) layer was then deposited onto MMT/PEI/substrate and washed using the same procedures described above, except using different rinsing containers. This deposition of a PEI layer, a PAA layer, and a MMT layered created a single TL (PAA/MMT/PEI/substrate). The procedure for depositing the next seven TLs was the same as the first TL, except the substrate was only soaked in the coating solutions for 1 min rather than 5 min. After the eight TLs were deposited, the specimen was dried in a convection oven (70 °C ± 1 °C, 12 h) and then stored in a dessicator (at least 3 d) with anhydrous calcium sulfate before weighing and analyzing. A more conventional MMT/PEI BL coating was also evaluated in this study. The process is identical to the TLs, except no PAA and 20 BL.

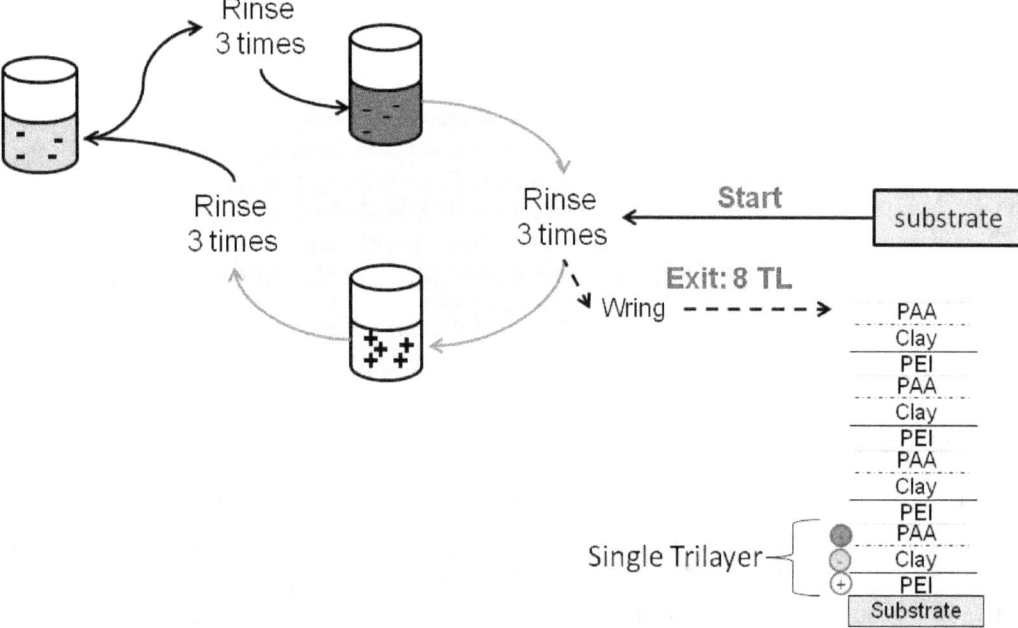

Figure 3. The MMT/polymer coating process was an alternating submersion in a polymer cationic (PEI), clay anionic (MMT), and polymeric anionic (PAA) solutions with washing (rinse and wring) between each solution. After creating five TLs (TL is PEI\MMT\PAA), the specimen was dried in a convection oven for 12 h at 70 °C ± 1 °C to remove excess water.

2.2. Fire performance testing

A Cone Calorimeter (Cone, Fire Testing Technology, East Grinstead, United Kingdom), operating at 35 kW/m² with an exhaust flow of 24 L/s, was used to measure the fire performance of uncoated and nanoparticle coated PUF. The experiments were conducted according to standard testing procedures (ASTM E-1354-07). A (10.2 cm / 10.2 cm / 5.1 cm) ± 0.1 cm sample was placed in a pan constructed from aluminum foil. The pan was slightly larger than the test sample and the pan sides were flared away from the sample. This allowed the sides as well as the top of the sample to be exposed during testing. When exposed to the external heat flux, the PUF and MMT/PUF melted to form a pool of burning polymer. In contrast, the MWCNT/PUF and CNF/PUF shrank slightly. As described by Zammarano et al. [53], it is critical to normalize the data according to surface because the PUF samples rapidly form a melt pool, whereas the, PUF filled with CNF and MWCNT did not collapse and maintained a two times larger surface area during the test. The data presented here was not normalized in terms of surface area since the surface areas for CNF/PUF and MWCNT/PUF were not quantitatively measured and changed significantly throughout the test.

All values are reported with a 2σ uncertainty of ± 5 % in HRR and ± 2 s in time. BF Cone testing was performed using the same methodology and sample size as PUF, except the BF was 0.4 cm ± 0.1 cm thick. There was no adjustment for surface area since there was no difference in the coated and uncoated BF dimensions during the test.

2.3. Simulated aging

A Head Over Heels shaker (HOH) was used to simulate the type of stresses expected from chewing on a piece of soft furnishing (Figure 4) [54]. A polymer coated glass bottle (300 mL) was charged with a simulated saliva solution (100 mL, 0.9 mass fraction % ± 0.08 mass fraction % sodium chloride in deionized water) and four substrates. The size of each substrate was a mass of 0.6 g ± 0.06 g with dimensions (length / width / thickness) of (4.0 / 5.0 / 1.0) cm ± 0.01 cm for the PUF and a diameter of 5.5 cm ± 0.01 cm for the BF. The extraction bottle was secured to a clamp positioned 8.0 cm ± 0.1 cm from the horizontal turning shaft of the HOH then spun at a rate of 6.3 rad/s (60 rev/min) for 30 min. The four substrates were removed after squeezing all the simulated saliva solution back into the bottle. Four new substrates were added to this bottle and the stressing and squeezing process was repeated. This process was repeated a total of 10 times; therefore, the simulated saliva extraction suspension to be analyzed by UV-VIS contained nanoparticles released from 40 PUF or BF specimens.

These suspensions contained sodium chloride, which was found to cause nanoparticle aggregation. In order to obtain accurate and reproducible measurements of the nanoparticles, sodium chloride was removed from the simulating chewing extracts using a Pierce Snakeskin pleated dialysis tube (10,000 g/mol relative molecular mass cut-off) suspended in deionized water. The dialysis tube was charged with approximately half of the extract (actual mass recorded) then the bag was placed in a deionized water bath (2 L, 0,2 μS), which was continuously agitated at room temperature. During every hour of the day the bath conductivity was measured and the water was replaced with fresh deionized water. At night, the dialysis proceeded unattended. In the morning of the following day, the conductivity was measured and the water replaced again. Throughout the day, the same process of hourly measuring conductivity and replacing the water continued until the measured conductivity was ≤ 0.5 μS (approximately 24 h), which was similar to the measured conductivity for the pure distilled water. As a comparison, the measured conductivity of the simulated saliva solution was 16,000 µS ± 300 µS. The dialysis tube was removed and the dialyzed simulated chewing extract was transferred to a beaker and the mass measured. The dialysis tubes were smaller than the volume of simulated chewing extract; therefore, the extract was dialyzed in two approximately equal mass parts. These dialyzed extracts were combined and then sodium dodecyl sulfate (0.23 mass fraction % ± 0.02 mass fraction %, SDS) was added to stabilize the nanoparticles in the suspension.

A mechanical pounder was used to simulate the type of "wear and tear" expected from routine use of soft furnishings (Figure 5). The "Pounder" was constructed of a hard plastic (diameter of 8.9 cm ± 0.1 cm) pounding element that had a slight convex curvature attached to a pneumatic driven vertical piston. The controller was a Trumeter 7922 counter and Trumeter 7951 timer. The substrate dimensions (length / width / thickness) for the pounding studies were (10.2 / 10.2 / 5.0) cm ± 0.1 cm for the PUF and the same surface dimensions for the BF, but the BF was only 0.4 cm ± 0.1 cm thick. The substrate was placed in a polyolefin bag then secured under the pounder by set screws positioned around the edge of the substrate. A substrate was pounded for 100,000 cycles (approximately 28 h) at 1 cycle/s at a force pressure of 20682 Pa ± 69 Pa (3 lbf/in^2 ± 0.1 lbf/in^2). After pounding was complete, the substrate was removed from the bag and the bag was washed with deionized water (100 mL) containing SDS (0.23 mass fraction % ± 0.02 mass fraction %) SDS. The same wash solution was used for 10 experiments; therefore, the 100 mL simulated wear and tear suspension contained the nanoparticles released from 10 substrates. The suspension was stored in a 300 mL glass bottle until analysis.

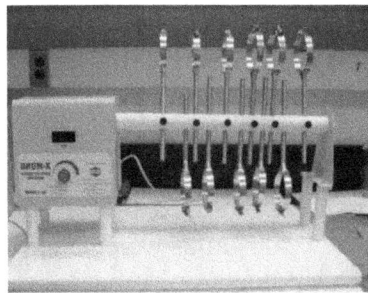

Figure 4. Simulated saliva extractions were performed using bottles containing a simulated saliva and a substrate that were rotated at 6.3 rad/s (60 rev/min) for 30 min.

Figure 5. Simulated normal wear and tear stressing was performed by pounding substrates at 1 cycle/s for 100,000 cycles at 20682 Pa ± 69 Pa of pressure.

2.2. Characterization

A Zeiss Ultra 60 Field Emission-Scanning Electron Microscope (FE-SEM, Carl Zeiss Inc., Thornwood, NY) was used to collect images of the nanocoatings, from which, the coating thickness was approximated (Table 1) and the distribution of nanoparticles and overall quality of the LbL coating was inspected. All SEM samples were sputter coated with 4 nm of Au/Pd (60 mass fraction %/40 mass fraction %) prior to SEM imaging.

A Q-500 GA Thermal Gravimetric Analyzer (TGA, TA Instruments, New Castle, DE) was used to measure the concentration of nanoparticles on the substrates (Table 1, Mass fraction % MWCNT on MWCNT/PUF). The samples (20 mg ± 3 mg) were placed on a ceramic pan (250 µL, TA Instruments) then loaded into the furnace by the autosampler. Under a nitrogen atmosphere, the temperature was stabilized at 90 °C ± 1 °C (30 min) then ramped to 800 °C ± 2 °C at 10 °C/min. The reported nanoparticle content was based on the remaining mass fraction % at 600 °C (CNF and MWCNT) and 800 °C (MMT). All values are reported with 2σ standard uncertainty.

A Lambda 950 Ultraviolet-Visible (UV-VIS, Perkin Elmer, Waltham, MA) spectrometer with a 10 mm ES quartz cuvette [55] was used to measure the mass fraction of CNF and MWCNT in suspensions generated from stressing the specimens. The absorbance spectrum was collected from 185 nm to 1800 nm at 1 nm increments with an instrument integration time of 0.2 s per increment. The incident light was circularly polarized prior to the sample compartment, and the instrument was corrected for both the dark current and the background. The relationship between the measured maximum absorbance band for the MWCNT and CNF (267 nm), after subtracting the spectrum of the simulate saliva/SDS solution, and the MWCNT or CNF concentration was determined from a Beer's Law calibration curve constructed from the absorbance of four different concentration calibration standards. The highest concentration calibration standard was 0.5 mg ± 0.01 mg of CNF or MWCNT suspended in SDS/DI (0.23 mass fraction % ± 0.02 mass fraction %) that was sonicated at 40 watts using a Sonics VCX130 sonicator with a 13 mm probe for 1h. The three lower concentration standards were aliquots of the highest concentration standard diluted with the same SDS/DI solution and sonicated for 10 min.

An Optima 5300 DV Inductively Coupled Plasma – Optical Emission spectrometer (ICP-OES, PerkinElmer Inc., Shelton, CT) was used to measure the mass fraction of MMT collected from stressing the clay coated substrates. The relationship between the measured ICP-OES intensity and MMT concentration in the stressed suspensions was determined from a series of Beer's Law calibration curves of four elements in the clay (Al, Fe, Mg, and Na). The intensity of Al, Fe, Mg, and Na in MMT was measured by selectively exciting each element at the element specific wavelengths (394 nm, 238.2 nm, 285 nm, and 589 nm, respectively). To determine the amount of each element in MMT, a 5 mL aliquot of a MMT/DI solution (0.020 mass fraction % ± 0.005 mass fraction % MMT, 100 mL) was first diluted with nitric acid (14 M, 100 mL) to give a MMT concentration of 0.0010 mass fraction % ± 0.0005 mass fraction % then known amounts of Al, Fe, Mg and Na were added and the ICP-OES intensity was measured. An additional amount of each element was added to this solution and the intensity was again measured. This process was repeated three more times and the ICP-OES intensities of each element in these five solutions were plotted as a function of concentration and element type. The concentration at an intensity value of 0 cps, is the amount of each element in MMT (0.0010 mass fraction % ± 0.0005 mass fraction %). The Beer's Law calibration curves were created by plotting the ICP-OES intensity of a 0.0010 mass fraction % ± 0.0005 mass fraction % MMT and three dilutions of this solution.

3. Results and Discussion

This project is investigating cutting edge technology that is based on decades of LbL research on flat, pseudo one dimensional substrates. Research, such as presented here, is required to understand how the parameters involved in fabricating these coatings will impact the quality and performance attributes of the substrate. To that point, the data provided here should be placed in the context of these coatings fabricated with a specific process, and type of nanoparticle, polymer, and substrate. All values are reported with a 2σ uncertainty.

SEM images of the as-received BF shows the "fiber bundle" surface is smooth (Figure 6) and contains micron sized grooves along the bundler axis. It is assumed that these grooves represent channels between fibers within the bundle not completely welded together during manufacturing. It appears that the 27 μm ± 3 μm bundles contained at least five fibers with a diameter of 9 μm ± 3 μm. The BF was not washed prior to coating because the non-woven construction was not durable enough to handle both prewashing and the coating process.

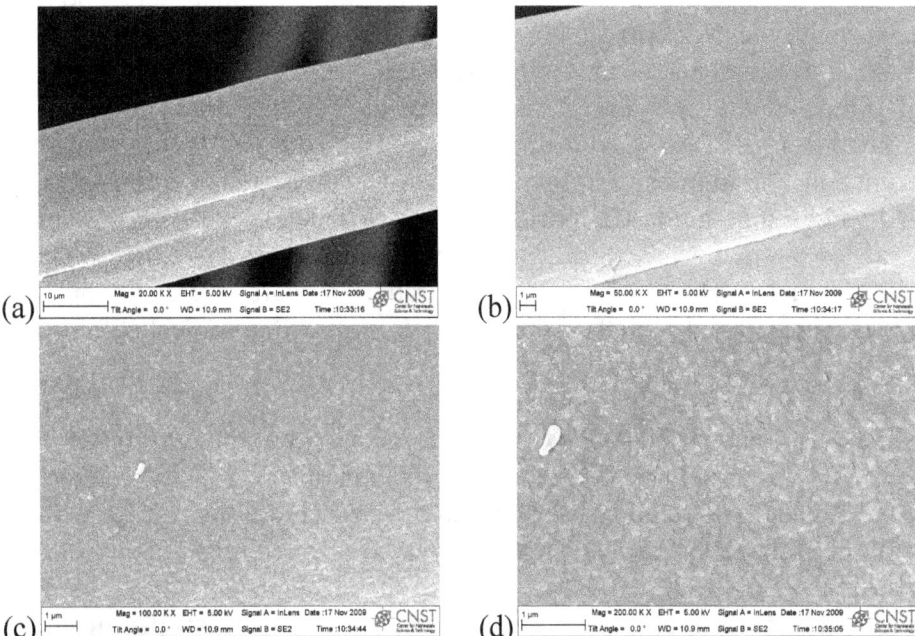

Figure 6. SEM images of as-received BF (a) 20,000x, (b) 50,000x, (c) 100,000x, and (d) 200,000x. The 27 μm ± 3 μm diamter strand is a bundle of several fibers that "welded"during manufacturing.

Other than dust and debris on the surface of the PUF, the as-received PUF surface appeared smooth and featureless even at high magnification (Figure 7). Prior to depositing the first layer, the PUF was washed with DI water, which completely removed all of the debris (Figure 7h). The wavy edges of the PUF walls were a result of the manufacturing process. The PUF was initially closed cell with a very thin polyurethane membrane connecting the walls. When the membrane was "popped" to form this open cell structure, there was a slight relaxation of the strained edges of the walls and the membrane snapped back onto the walls, which created the wavy appearance observed in Figure 7h.

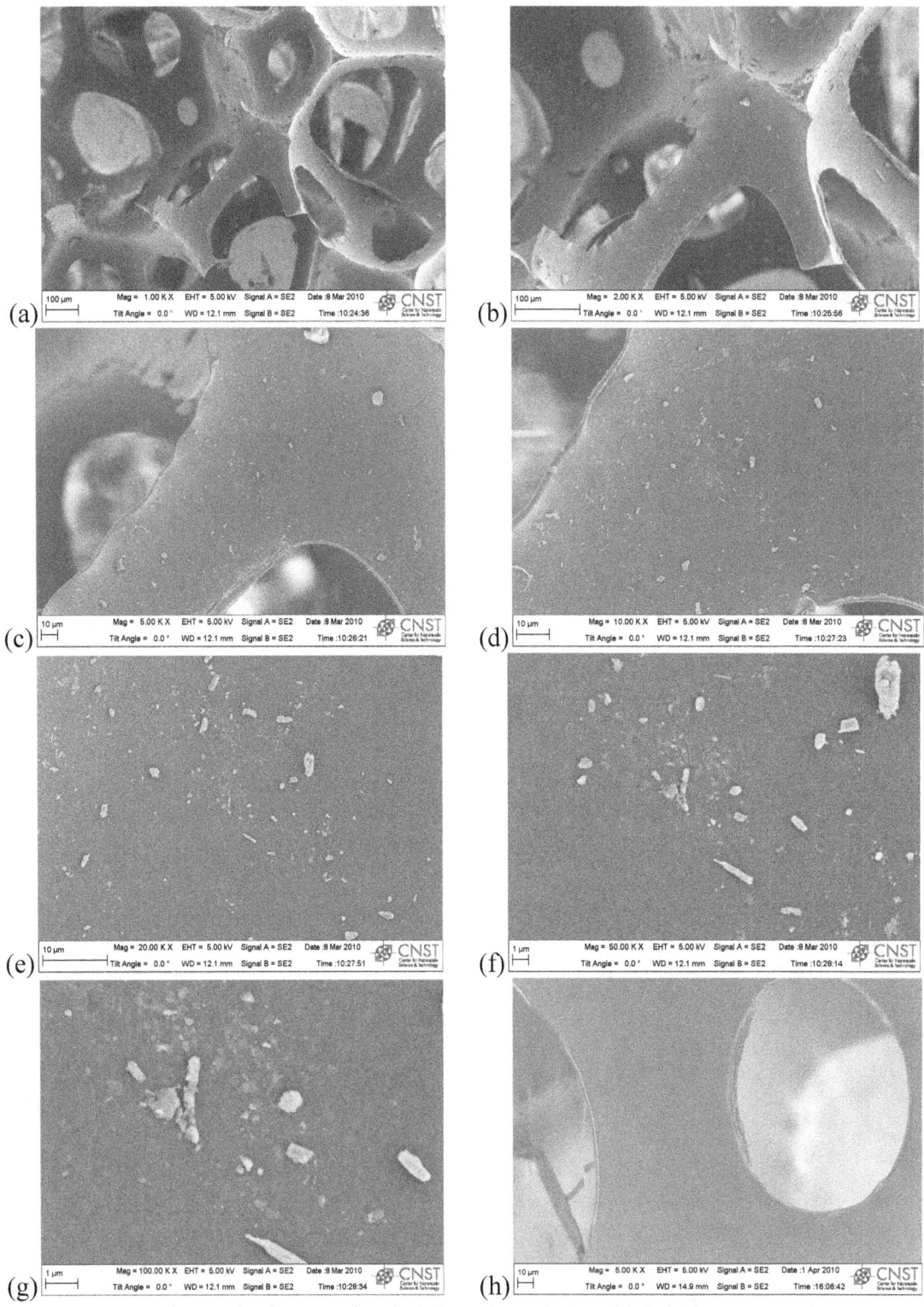

Figure 7. SEM images of as-received PUF at (a) 1,000x, (b) 2,000x, (c) 5,000x, (d) 10,000x, (e) 20,000x, (f) 50,000x and (g) 100,000x and washed PUF at (h) 5,000x. The PUF surface was smooth and featureless after debris (dust, etc.) was removed by washing (h).

3.1. Characterization of nanocoatings

The physical characteristics of the nanoparticle coated substrates are provided in Table 1. The increase in substrate mass due to the coating (Mass fraction % coating) was measured using a laboratory microbalance. The amount of nanoparticles in the coating (Mass fraction % Nanoparticles in coating) was calculated from TGA and microbalance values. The four BL CNF coatings of PUF increases the mass of the substrate by 3.2 mass fraction % ± 0.4 mass fraction %, of which, 51 mass fraction % ± 3 mass fraction % is CNFs. The total CNF content relative to PUF mass is 1.6 mass fraction % ± 0.1 mass fraction %. The physical characteristics of the four TL MWCNT coatings on PUF (increase the mass of the substrate by 3.4 mass fraction % ± 0.4 mass fraction %, of which, 50 mass fraction % ± 3 mass fraction % is MWCNTs) are similar to that of the CNF coatings. The total MWCNT content relative to the substrate mass is 1.7 mass fraction % ± 0.1 mass fraction %. The MMT coatings on PUF are significantly thicker (1000 nm ± 450 nm) with a higher nanoparticle concentration in the coating and on the substrate (66 mass fraction % ± 13 mass fraction % and 2.1 mass fraction % ± 0.2 mass fraction %, respectively). These nanoparticle loading levels on the substrate were similar to what is used to improve the fire performance of polymers [27,31]. In contrast to these more conventional nanocomposites, the nanoparticles in these coatings were concentrated at the surface rather than randomly dispersed and distributed throughout the polymer matrix

The characteristics measured for the BFs are similar to what was measured for the PUF specimens of the same nanoparticle-based coatings, except for MMT. For example, the CNF content in the coatings on PUF and BF are 50 % ± 3 %. Since the MMT coating process deposited 2.2 times more coating (by mass fraction) on the BF than the PUF, the thickness of the coating on the BF is presumably thicker. The freeze fracture process used to measure the coating thickness on the PUF did not work on BF. Therefore, the coating thickness values on PUF are used as an estimate of the BF values for the MWCNT and CNF. Justification for this assumption is that both substrates were coated using the same processes and the TGA and microbalance values indicate both substrates contained the same mass of nanoparticles and coating.

Table 1. Provided are the average physical characteristics of nanoparticle coated substrates organized by substrate type and highest to lowest in nanoparticle content. BFs coating thickness is assumed to be similar to that measured on PUF. MMT coatings are sufficiently different on PUF and BF that the coating thickness on PUF is believed to not be a good estimate of the BF coating thickness. Values reported with 2σ uncertainty.

Specimens	Specimen Mass (g)	Mass fraction % coating	Mass fraction % Nanoparticles		Coating thickness (nm)
			on substrate	in coating	
PUF					
MMT	15.9 ± 0.3	3.2 ± 0.6	**2.1 ± 0.2**	66 ± 13	1000 ± 450
MWCNT	13.2 ± 0.3	3.4 ± 0.4	**1.7 ± 0.1**	50 ± 3	440 ± 47
CNF	13.1 ± 0.3	3.2 ± 0.4	**1.6 ± 0.1**	51 ± 3	359 ± 36
BF					
MMT	5.5 ± 0.5	6.9 ± 1.4	**4.6 ± 1.0**	66 ± 15	---
MWCNT	5.0 ± 0.2	3.6 ± 0.5	**1.8 ± 0.3**	51 ± 3	440 ± 47
CNF	5.5 ± 0.4	3.5 ± 0.3	**1.7 ± 0.2**	49 ± 3	359 ± 36

3.1.1. CNF-based nanocoatings

The images in Figure 8 indicate that the CNFs are well distributed along the walls of the PUF. At low magnification (Figure 8a), the wall surfaces appear to be sparsely populated, with approximately 10 μm by 10 μm sized aggregates of CNFs. The areas between the aggregates are populated with a network of CNF whiskers and regions that appear to be free of CNFs. At higher magnifications (Figure 8c to Figure 8g), it becomes apparent that a portion of these regions actually do contain CNFs and these are not visible at lower magnifications because they are completely embedded in the polymer coating. Along the surface there are also "islands" of CNFs that appear to have dewetted from the surface (Figure 8d to Figure 8g). The larger islands (approximately 10 μm) appear to contain more single CNFs, rather than the bundles observed in the larger aggregates (Figure 8h). The smaller islands (less than 3 μm) either contain no CNFs or what appears to be short individual CNFs (less than 1 μm).

SEM images of a fractured CNF/PUF were taken with the fracture surface in the plane of the image, which provides cross section views of the PUF and the coating (Figure 9). The CNF coatings are 359 nm ± 36 nm (based on 10 measurements of five different CNF/PUF specimens. The surface morphology at low magnification (Figure 9a) is consistent with that observed in Figure 8 (large aggregates, CNF network, and areas without CNF). Based on all the images taken of fractured CNF/PUFs, the CNF coating appears to cover the entire surface, although the thickness is not completely uniform.

The thicker islands, one of which is shown in Figure 9c, are 374 nm ± 100 nm and are constructed of at least twenty fibers randomly oriented in the plane of the coating. The "hills and valleys" topography of the islands are created by overlapping CNFs. The fairly uniform (34 nm ± 2 nm) polymer coating covering the CNFs, and the small gaps between them, suggest that these fibers were probably deposited at the same time. Analysis of other islands reveals similar characteristics, which suggests the fibers in the islands were probably deposited as loosely interacting groups rather than isolated and independent CNFs. The last layer of polymer covering the CNFs varied from island to island, but never exceeded 34 nm ± 2 nm. This is the same thickness measured for the regions between the thicker islands, which appear to contain little to no CNFs. Figure 10 is the only example of coating delamination. The delamination may have resulted from the freeze fracture process or may indicate a section of poor adhesion. These images illustrate that the coating can contain regions of high CNF concentration welded together by polymer.

The SEM images of the CNF/BFs indicate the CNF coatings are similar on PUF and BF (Figure 11). More specifically, there are regions of high fiber aggregation separated by regions with little (or no) CNFs and islands that contain high CNF aggregation or individual CNFs. The entire BF surface is coated with this non-uniform CNF distribution.

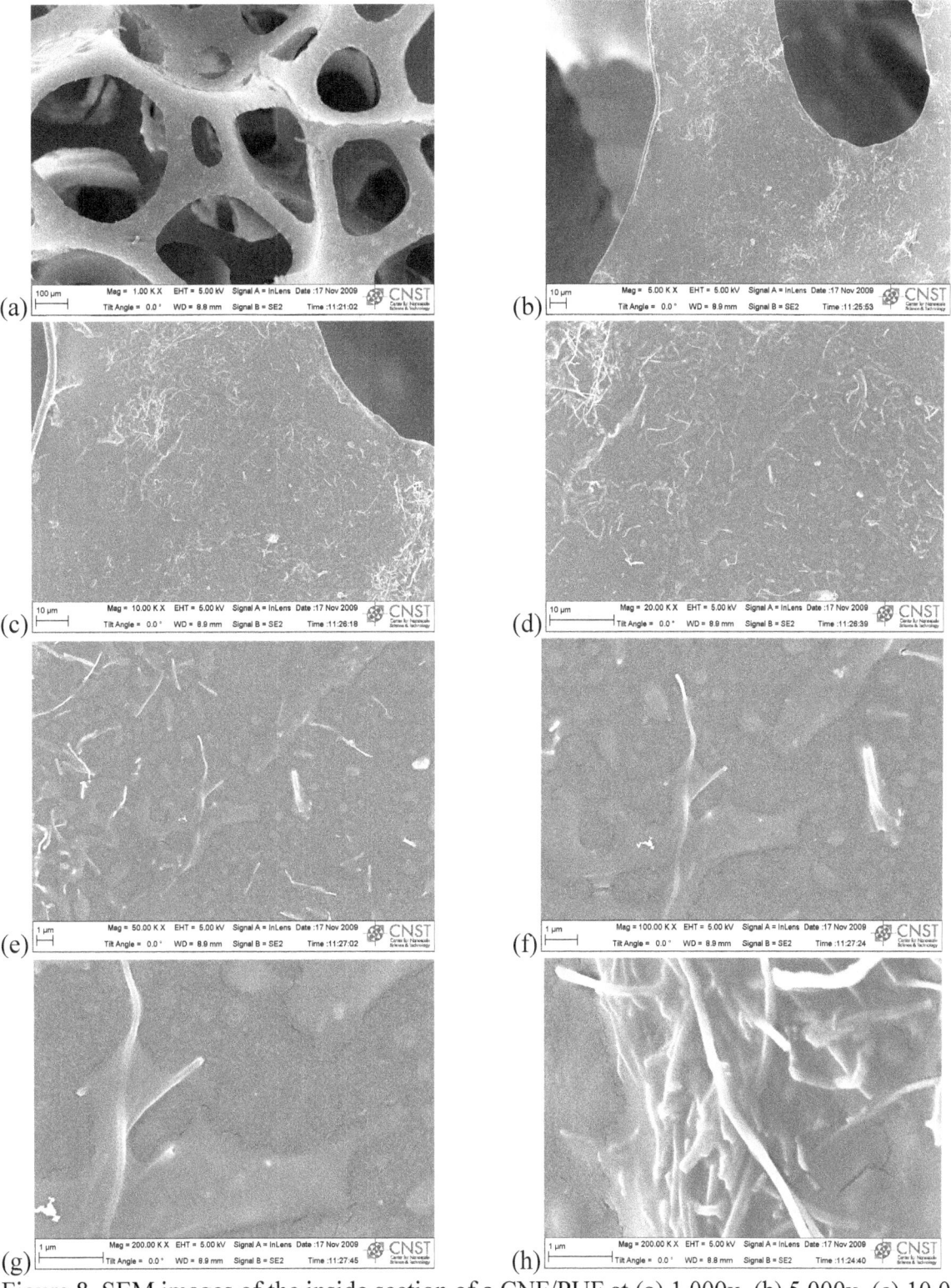

Figure 8. SEM images of the inside section of a CNF/PUF at (a) 1,000x, (b) 5,000x, (c) 10,000x, and (d) 20,000x, of a thicker island at (e) 50,000x (f) 100,000x, and (g) 200,000x, and of an aggregate at (h) 200,000x.

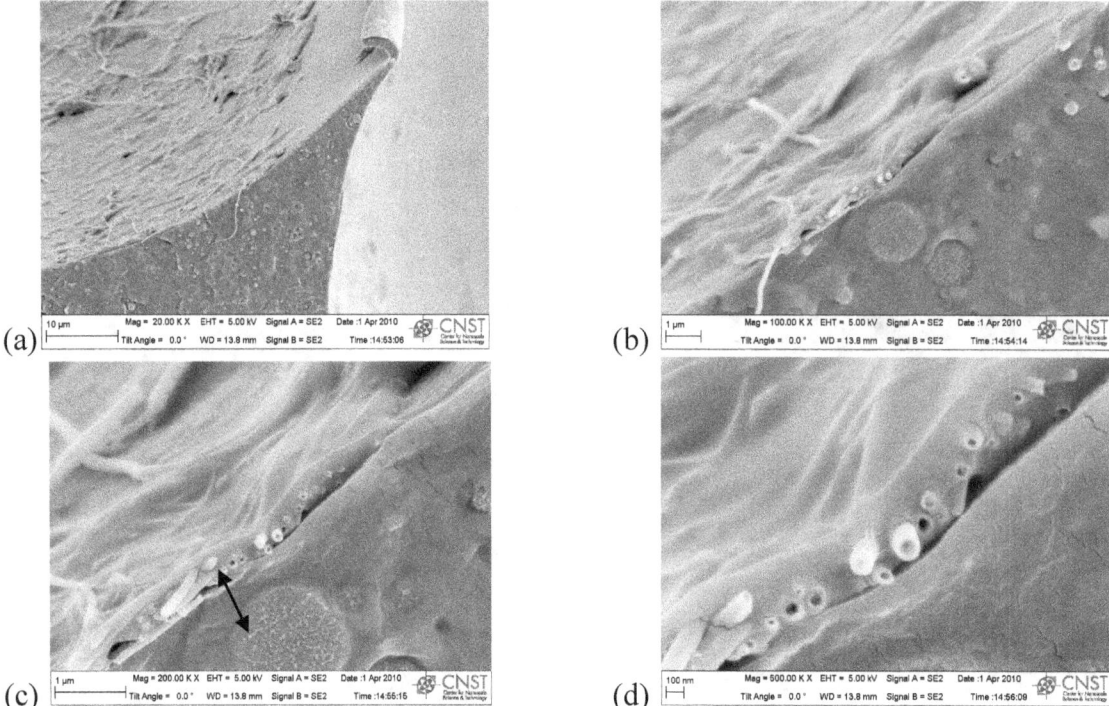

Figure 9. SEM images of a fractured edge of CNF/PUF at (a) 20,000x, (b) 100,000x, (c) 200,000x, and (d) 500,000x. The CNF coating is 359 nm ± 36 nm.

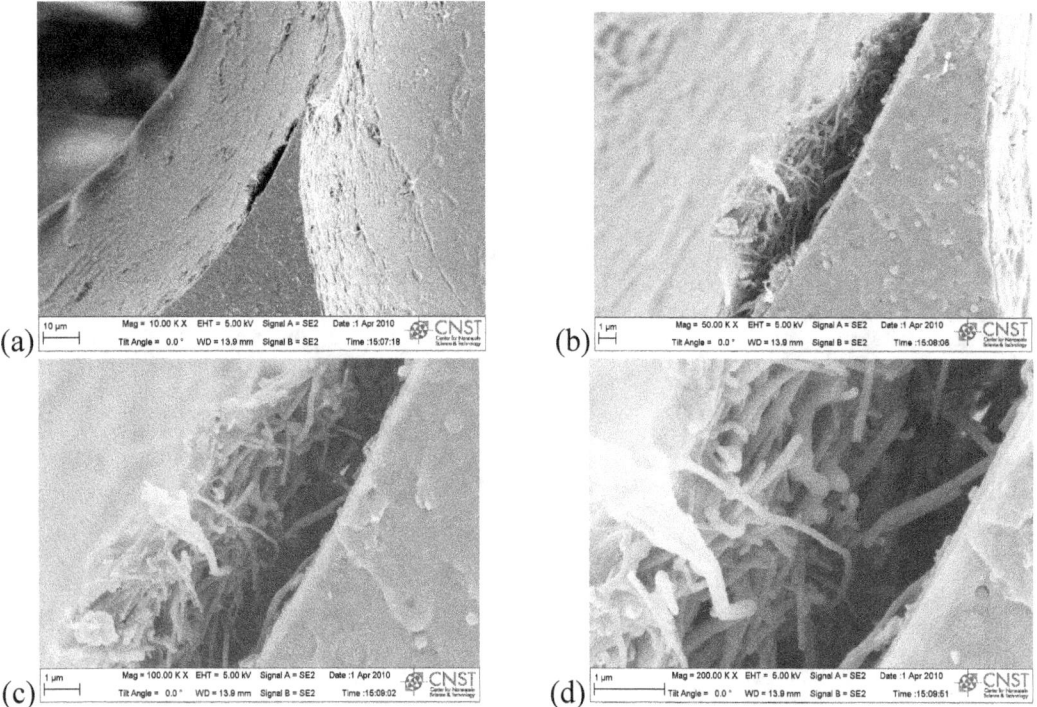

Figure 10. SEM images of a delaminated CNF/PUF at (a) 10,000x, (b) 50,000x, (c) 100,000x, and (d) 200,000x. The CNFs below the surface are welded together with polymer. The root cause of delimitation may be the freeze fracture process or poor adhesion to the PUF due to the high CNF concentration.

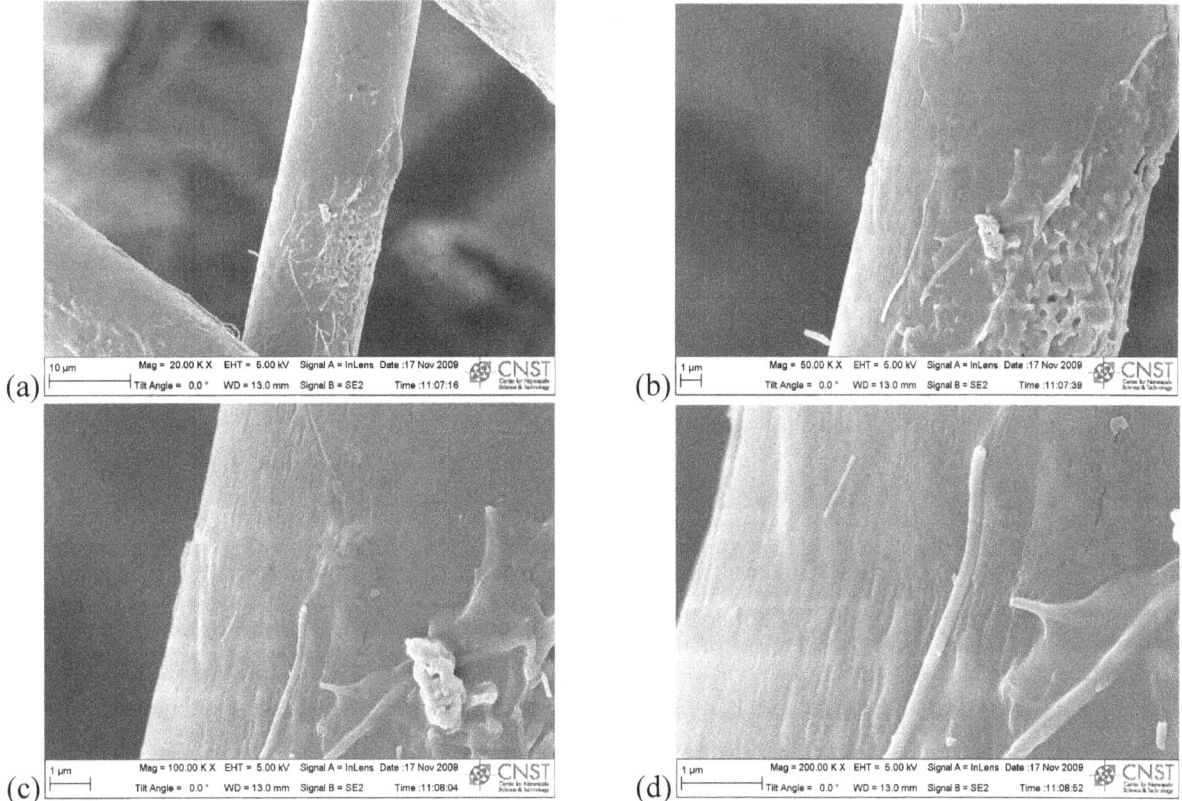

Figure 11. SEM images of the inside section of a CNF/BF at (a) 20,000x, (b) 50,000x, (c) 100,000x, and (d) 200,000x and (e) 200,000x. The CNF coatings on the PUFs and BFs appear to be similar.

3.1.2. MWCNT-based nanocoatings

Uniform coatings with well dispersed and distributed MWCNTs were only achieved when the MWCNTs were well stabilized in the MWCNT depositing solution over the entire coating process. Critical to fabricating the MWCNT coatings is to functionalize the MWCNTs with PEI using a procedure by Liao [52] and to use TL, rather than BL, coatings. PEI and PAA were selected for these coatings because their well documented LbL behavior makes them an idea starting point for evaluating new additives and substrates for the LbL process.

Initially, a stable MWCNT suspension was attempted using PEI or sodium deoxycholate surfactant. After sonicating for one hour, the MWCNTs appeared well dispersed and stable; however, for sonication over an hour, the MWCNTs appeared to destabilize as the bottom of the suspension was darker than the top. A couple of the more stable suspensions were used to fabricate a coating. Even though the PUF was washed well between depositions, because it is a three dimensional porous substrate, there was still unbound PAA present in the PUF when the MWCNTs were deposited. This PAA caused the MWCNT suspension to rapidly destabilize as indicated by a settling of the MWCNTs on the bottom of the suspension. The result from these attempts was a coating with a very non-uniform distribution of MWCNTs.

A drastic improvement in the process and therefore the coatings resulted from using MWCNTs functionalized with PEI. The MWCNT-PEI formed a stable suspension in DI over several days without the need for a stabilizing surfactant. The high dispersion and distribution of MWCNTs and the high stability of the suspension was indicated by the very uniform black color (caused by MWCNT-PEIs) across the PUF. However, the black MWCNT-PEIs were easily transferred from the PUF into the PAA and, to a lesser extent, to the rinsing solutions. This suggested that the MWCNT-PEIs had a relatively weak charge, as compared to PAA, which resulted in poor adhesion between these layers. The retention was addressed by depositing a PEI layer after the MWCNT-PEI (creating a TL) as the adhesion was then based more on pure PAA and PEI interactions, which strongly adhere to each other.

At low SEM magnifications, the LbL process appears unsuccessful (Figure 12a through Figure 12e). The MWCNT/PUF surface appears void of MWCNTs other than a few sparsely populated MWCNT aggregates that are on the order of tens of micron in size. However, images at higher magnifications (Figure 12e through Figure 12h) revealed that the LbL process had worked extremely well. The PUF surface was completely covered with a uniform coating, which contained well dispersed and distributed MWCNTs that are completely embedded in the polymer coating. All of the MWCNT/PUFs analyzed by SEM had a completely intact MWCNT protective layer over the PUF. A few small surface cracks (10 nm ± 5 nm) were observed and were believed to result from the drying process. These types of cracks were common in thicker LbL coatings. It was assumed that these rare nanometer scale surface cracks would not deteriorate the fire performance of the MWCNT/PUF; therefore, there was no further investigation of the cracks.

SEM images of a fractured MWCNT/PUF were taken with the fracture surface in the plane of the image, which provides cross sectional views of the PUF and the coating (Figure 13). The MWCNT coating was 440 nm ± 47 nm thick based on seven measurements taken on each of 12 different MWCNT/PUF specimens. Other than the MWCNTs exposed at the fracture surface, all MWCNTs were completely embedded in the polymer coating. The cracks in the MWCNT coating were likely there prior to fracturing, but the size was increased by the strain resulting from the freeze fracture process.

The SEM images of the MWCNT/BFs indicate the MWCNT coatings were similar on PUF and BF (Figure 14). More specifically, the entire surface of the BF was coated with uniform distribution and high concentration of MWCNT.

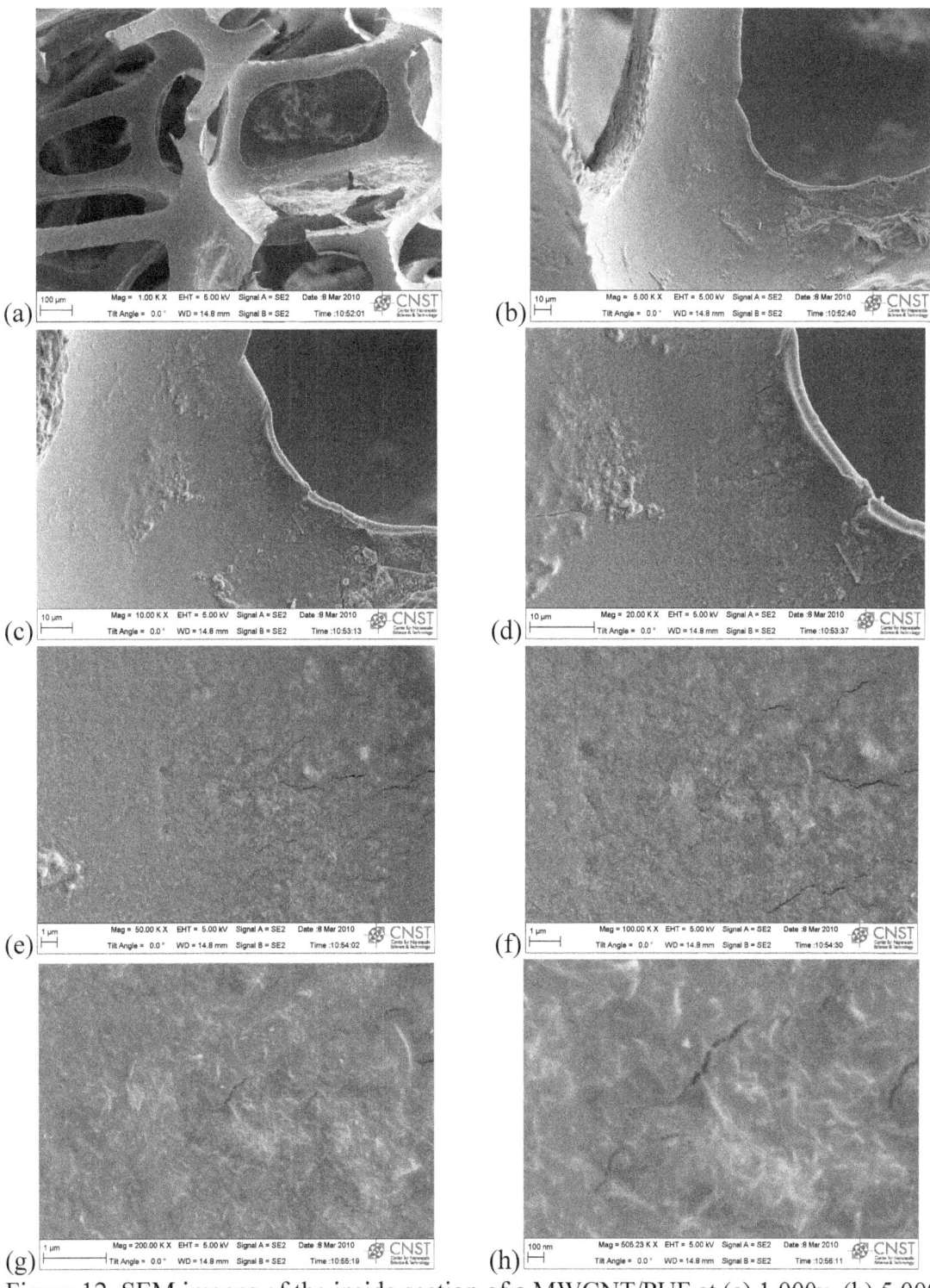

Figure 12. SEM images of the inside section of a MWCNT/PUF at (a) 1,000x, (b) 5,000x, (c) 10,000x, (d) 20,000x, (e) 50,000x (f) 100,000x, (g) 200,000x, and (h) 500,000x. The MWCNTs were well dispersed and distributed throughout the polymer coating. The coating was smooth and featureless except for a few small larger aggregates and a few 10 nm ± 5 nm wide cracks.

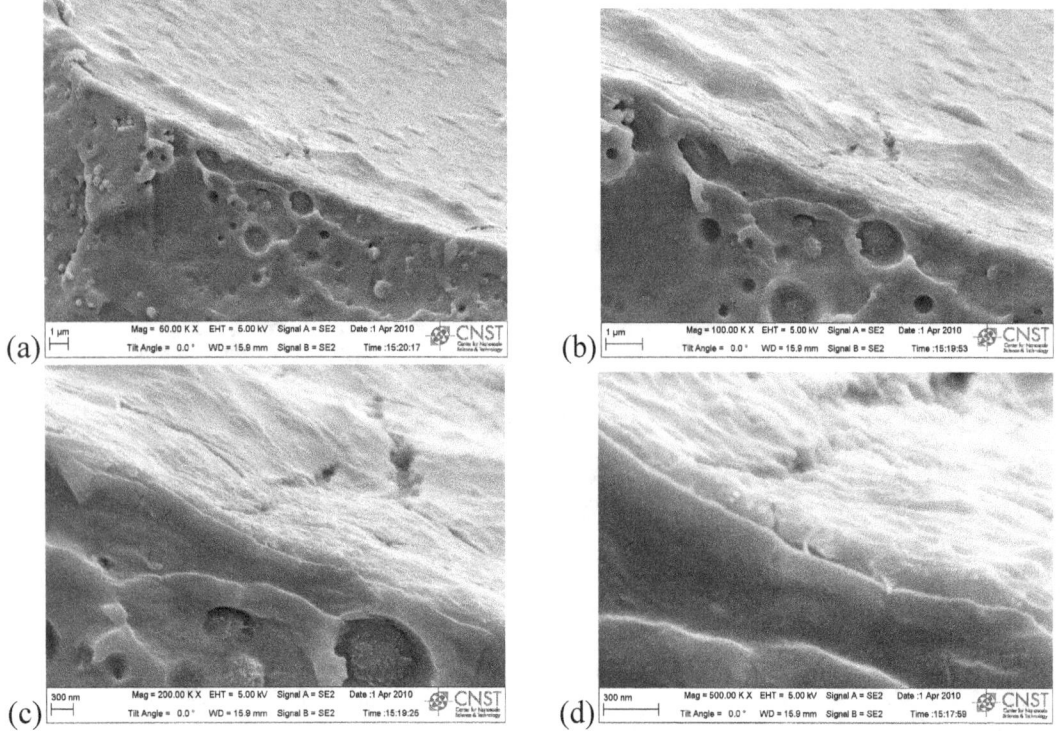

Figure 13. SEM images of a fractured edge of MWCNT/PUF at (a) 50,000x, (b) 100,000x, (c) 200,000x, and (d) 500,000x. The MWCNT coating thickness is 440 nm ± 47 nm.

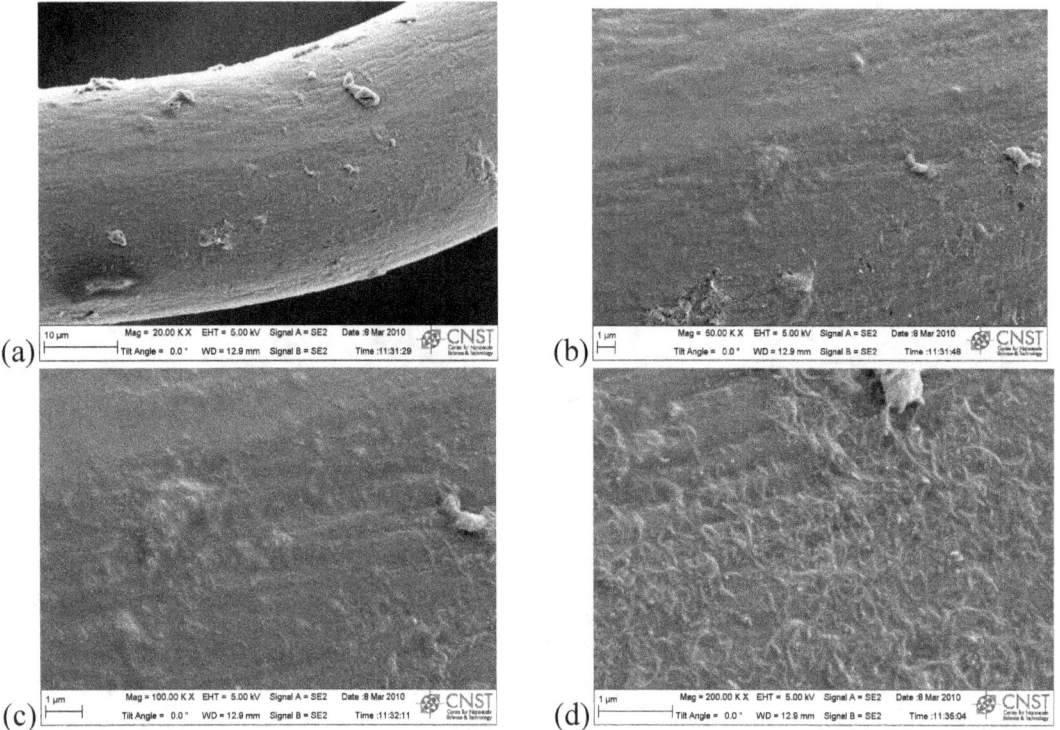

Figure 14. SEM images of the inside section of a MWCNT/BF (a) 20,000x, (b) 50,000x, (c) 100,000x, and (d) 200,000x. The MWCNT coatings appear similar on PUF and BF.

3.1.2. MMT-based nanocoatings

The images in Figure 15 indicated the entire PUF surface was coated with a non-uniform distribution of clay. The surface was covered with regions of high MMT aggregation that can be as large as 100 μm by 100 μm. Fracture surface images (Figure 16) indicated these regions can be several microns thick. The coating thickness was 1000 nm ± 450 nm. The large uncertainty in the thickness stems from a large variation in the degree of clay aggregation, which can be seen in Figure 16b. The entire left wall in image Figure 16b is covered with a high aggregation of MMT with a thickness of 1250 nm ±150 nm. On the top wall, the coating is 1000 nm ± 100 nm on the left, but thins out to 450 nm ± 56 nm on the right. On the right wall, the coating appears to be very similar in thickness to the top right, but the part of this wall in the background has the same rough, highly aggregated appearance that is observed on the left wall. This disparity is also shown in several images in Figure 16 and Figure 15. The smooth and featureless regions between these large aggregates (Figure 15a and Figure 15b) are actually completely filled with clay (Figure 15c to Figure 15f). The coating thickness in these regions is closer to what was expected from this process (500 nm ± 120 nm). A cross section of a thickness average region of the coating (1000 nm) indicates the coating is highly filled with MMT sheets stacked upon each other similar to a deck of cards (Figure 16c and Figure 16d).

The SEM images of the MMT/BFs indicate that the MMT coatings were similar on PUF and BF (Figure 17), except there were more regions with highly aggregated clay on the BF. The large variation in aggregate size is well illustrated in these BF images. On the front of the bundle, the MMT aggregates appear as flakes that could easily be peeled away. In contrast, the top is covered with both very tall and completely embedded clay aggregates and smooth, thinner coated regions (similar to what is presented on PUF in Figure 15 and Figure 16). The root cause of the variation in aggregation, regions of extremely thick and highly aggregated clay, the high clay concentration (relative to the nanoparticles in the other coatings), and the higher coating mass on BFs (relative to PUF), is currently under investigation. It is assumed using a TL rather than the more traditional BL approach contributed to these characteristics.

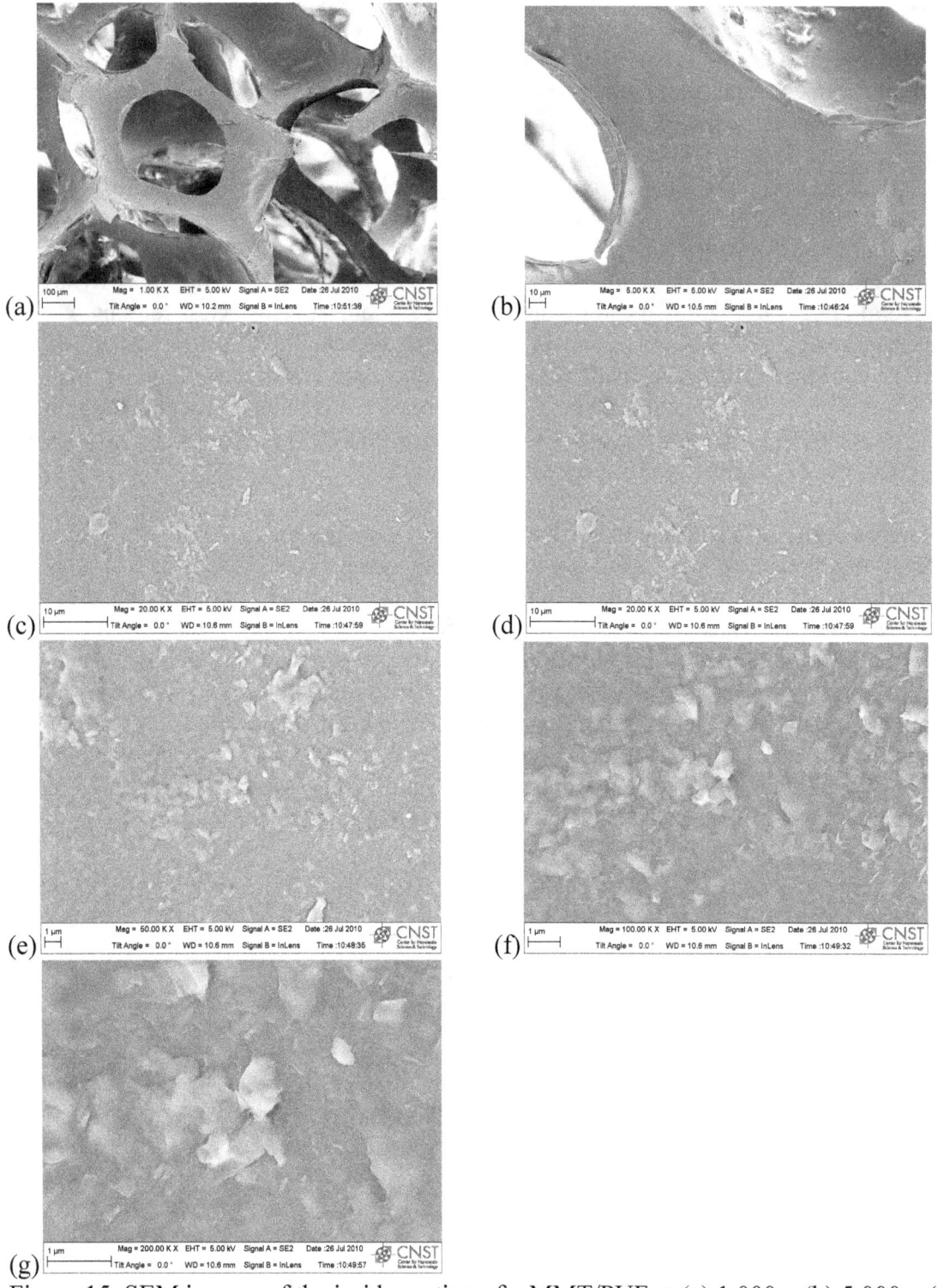

Figure 15. SEM images of the inside section of a MMT/PUF at (a) 1,000x, (b) 5,000x, (c) 10,000x, (d) 20,000x, (e) 50,000x (f) 100,000x, and (g) 200,000x.

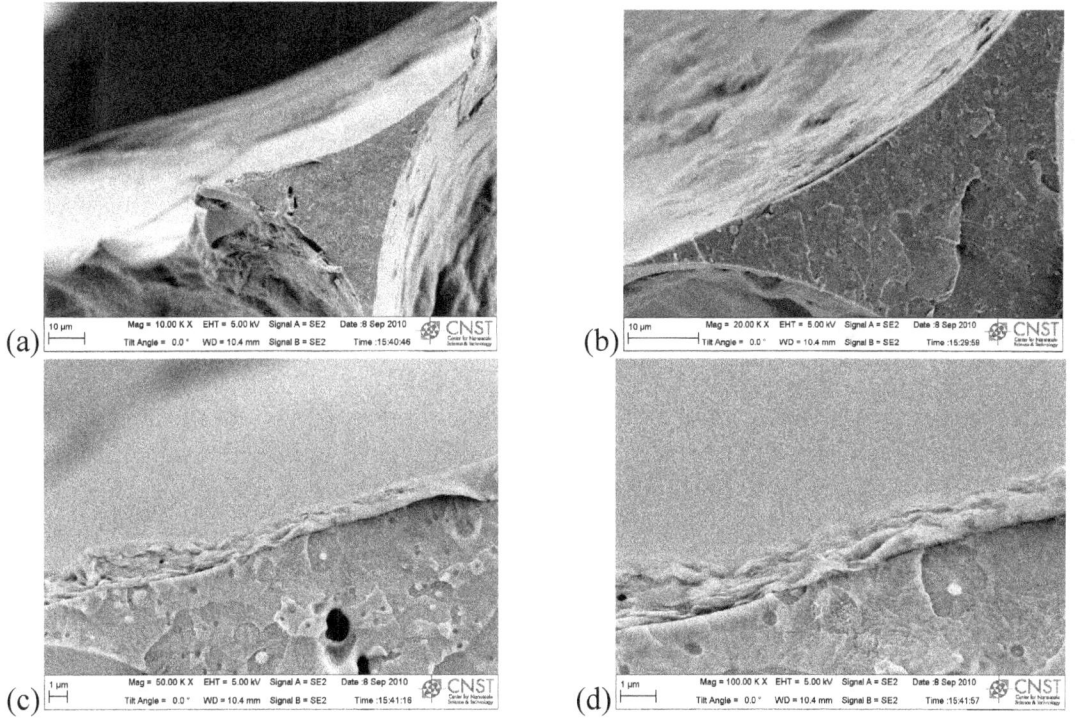

Figure 16. SEM images of a fractured edge of MMT/PUF at (a) 10,000x, (b) 20,000x, (c) 50,000x, and (d) 100,000x. The MMT coating was 1000 nm ± 450 nm.

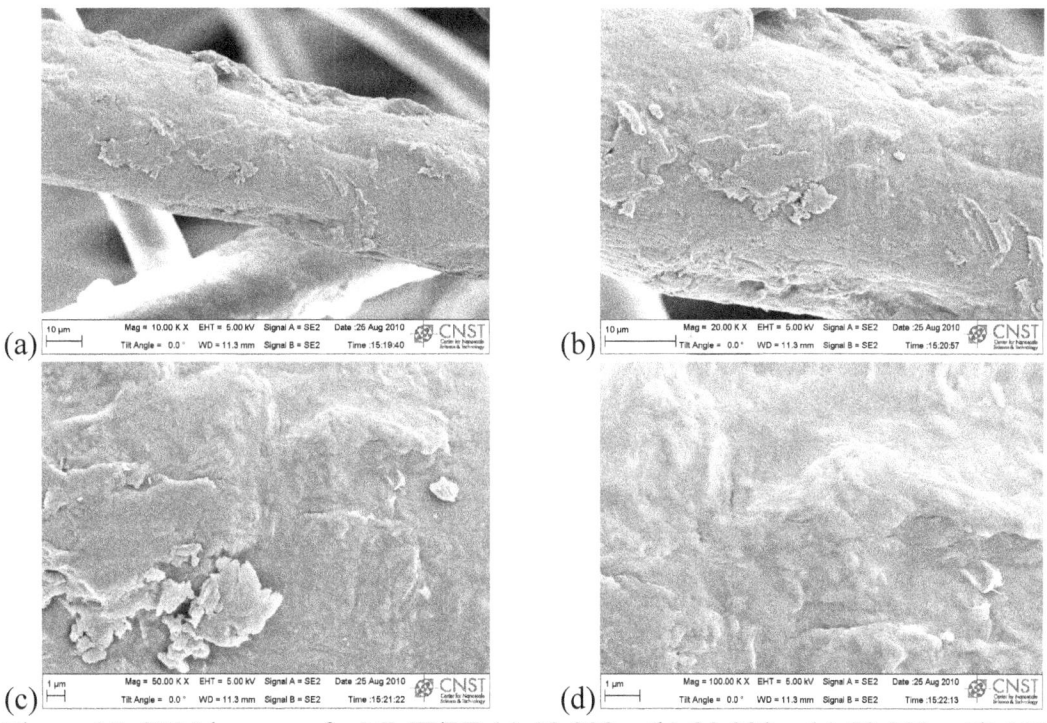

Figure 17. SEM images of a MMT/BF (a) 10,000x, (b) 20,000x, (c) 50,000x, (d) 100,000x and (e) 200,000x. The MMT coatings are on PUF and BF.

3.2. Fire performance

The Cone Calorimeter is a routine bench scale fire test that simulates a developing fire scenario on a small specimen and is used to measure the forced burning fire performance of polymers. The parameters reported from the test, such as time to ignition of the combustion gases (TTI), the time to peak, the peak maximum heat release rate (PHRR), and the total heat release (THR), are directly related to the potential fire threat of the burning polymer and, therefore, the values of these parameters are the bases of the performance metrics for several existing or proposed national fire regulations.

3.2.1. CNF-based nanocoatings

TGA was used to determine the actual mass of CNFs and coating deposited onto the substrates. The four BL CNF coatings increase the mass of the PUF and BF substrates by 3.4 mass fraction % ± 0.1 mass fraction %, of which, 50 mass fraction % ± 4 mass fraction % are CNFs. The total CNF content relative to the substrate mass is 1.6 mass fraction % ± 0.3 mass fraction %, which is a typical loading level of carbon nanotubes or nanofibers incorporated (embedded) into polymers to improve the polymer's fire performance. Unlike these other nanoparticle embedded nanocomposites, the CNFs in these coatings are concentrated at the surface rather than randomly dispersed and distributed throughout the polymer matrix [29,31].

The HRR data indicates the CNF coatings significantly improved the fire performance of foam (Figure 18 and Table 2). The HRR curves for CNF/PUF and PUF consist of two peaks. However, the attributes of the curves are quite different with both peaks of CNF/PUF being of similar HRR values (371 kW/m^2 ± 10 kW/m^2 and 348 kW/m^2 ± 10 kW/m^2) whereas in PUF the second peak is 2.8 times larger than the first peak (224 kW/m^2 ± 12 kW/m^2 and 620 kW/m^2 ± 26 kW/m^2). The PHRR, which is often considered as one of the more critical values in accessing the flammability of a material, is 40 % ± 3 % lower for CNF/PUF than for PUF. The THR, which reflects the total size of the fire threat, and the total burning time of the foam was 21 % ± 3 % smaller for CNF/PUF. However, the time to PHRR, which is often considered a critical value in accessing the amount of time for escaping a fire, is 66 % ± 2 % earlier for CNF/PUF. In other words, the Cone data suggests the CNF coatings may result in smaller fires and reduced fire spread, but create an initially larger fire that reduces the time to escape. In a real fire scenario, the CNF/PUF would likely perform significantly better than the HRR data suggests, but before going into this discussion it important to understand how the CNF coating altered the burning behavior of foam.

The attributes of the first HRR peak for both CNF/PUF and PUF are defined by pyrolysis of polyurethane decomposition gases (increase in HRR) and decrease in substrate surface area (decrease in HRR) (Figure 18). During the first peak the CNF coating forms a protective char and enables the foam to maintain its shape (and surface area) that is qualitatively similar to an untested foam specimen. At 63 s ± 2 s, the flames penetrate this protective char, which causes the foam to shrink as the remaining polymer is pyrolyzed. At the end of the experiment, a brittle char remains that has a surface area 90 % ± 5 % smaller than the untested foam specimen. In contrast to CNF/PUF, during the first HRR peak PUF collapses to form a liquid-like pool of degraded polyurethane. The surface area of this pool (defined by the sample pan) is qualitatively two times smaller than CNF/PUF surface area, which is the reason the HRR maximum for the first peak is 66% ± 2% higher for CNF/PUF (Table 2). The PUF pool rapidly pyrolyzes because there is no protective char and because the contents of the pool are volatile/combustible compounds (isocyanate and polyol based degradation products of polyurethane).

In a real fire scenario, the reduction in flammability due to replacing PUF with CNF/PUF will likely be greater than suggested by the Cone data (Figure 18 and Table 2) for two main reasons. First, this CNF/PUF Cone data (Figure 18 and Table 2) was not normalized, as suggested by Zammarano et al [53], since the surface area for CNF/PUF was not quantitatively measured and changed significantly throughout the test. Based on the qualitative observations, the HRR for the first peak could be reduced by a factor of two while the second peak may only be slightly reduced. The result is the first peak for CNF/PUF would have a maximum HRR and time to peak similar to PUF and the second peak would then become the PHRR, which has a time to peak similar to PUF. Secondly, the Cone data does not capture the real impact of the PUF pool fire. Since a pool fire can approximately increase the fire threat (as calculated from HRR, THR, and burn time) of a burning product (e.g. upholstered furniture) by 35 % [59] and the CNF coating prevents pool formation, it is hypothesized that in a real fire replacing PUF with CNF/PUF would decrease the HRR from a product by at least 35 %.

The CNF coating on the BF increases the BF flammability (Figure 18 and Table 3). More specifically, a 2.6 times increase in PHRR (149 kJ/m^2 ± 15 kJ/m^2 as compared to 57 kJ/m^2 ± 9 kJ/m^2) and a 3.9 times increase in THR (3.1 MJ/m^2 ± 0.3 MJ/m^2 as compared to 0.8 MJ/m^2 ± 0.1 MJ/m^2) was measured for the CNF/BF as compared to pure BF (Figure 19 and Table 3). This increase is attributed to the coating polymers being more flammable then the BF itself and the coating process causing the BF to fall apart. The non-woven design is not appropriate for this process because depositing, washing, and rinsing creates regions of low fabric density, loss of fibers, and collapse of the air pocket design.

3.2.2. MWCNT-based nanocoatings

TGA was used to determine the actual mass of MWCNTs and coating deposited onto the substrates. The five TL MWCNT coatings increase the mass of the PUF and BF substrates by 3.5 mass fraction % ± 0.5 mass fraction %, of which, 50 mass fraction % ± 3 mass fraction % is MWCNTs. The total MWCNT content relative to the substrate mass is 1.8 mass fraction % ± 0.4 mass fraction %, which is a typical loading level of carbon nanotubes or nanofibers incorporated (embedded) into polymers to improve the polymer's fire performance. However, unlike these other nanoparticle embedded nanocomposites, the MWCNTs in these coatings are concentrated at the surface rather than randomly dispersed and distributed throughout the polymer matrix [29,31].

The HRR data indicated that the MWCNT coatings significantly improved the fire performance of foam (Figure 18 and Table 2). The HRR curves for MWCNT/PUF and PUF consist of two peaks. The time to maximum HRR for each peak was 21% ± 10 % earlier for PUF (32 s ± 2 s and 75 s ± 3 s as compared to 34 s ± 2 s and 78 s ± 2 s for MWCNT/PUF). The maximum measured HRR during the test (PHRR) was 35 % ± 3 % lower for MWCNT/PUF (620 kW/m^2 ± 26 kW/m^2 as compared to 403 kW/m^2 ± 10 kW/m^2). The THR and total burn time was also lower for MWCNT/PUF (21 % ± 3 % and 25 % ± 3 %, respectively).

The improvement in fire performance of the MWCNT/PUF was a result of foam shape retention and char formation caused by the MWCNT coating network. Similar to CNF/PUF, the attributes of the first peak for both MWCNT/PUF and PUF were defined by an increase in HRR due to pyrolysis of polyurethane decomposition gases and a decrease in HRR due to collapse (PUF) or shrinkage (MWCNT/PUF) of the foam and char formation (MWCNT/PUF). At (45 ± 2) s, the PUF had completely collapsed into a pool of pyrolyzing polyurethane foam decomposition products with a surface area equivalent to the dimensions of the sample pan (10 cm by 10 cm). The subsequent rapid HRR increase for PUF (that resulted in the second peak and the PHRR) was a result of pyrolyzing the degradation products in this pool. Unlike PUF, during the first peak the MWCNT/PUF only shrank. Qualitatively, the MWCNT/PUF was comparable in dimensions to the untested foam, which was a surface area two times larger than PUF. At (54 ± 2) s, the flames penetrated the protective MWCNT based char layer allowing for the remaining polyurethane to pyrolyze. The result was a second HRR peak, significant shrinkage (10% ± 5% of the original surface area by the end of the test), and brittle char residue.

Similar to CNF/PUF, a potential concern of the MWCNT coating was the 75 % ± 2 % higher HRR for the first peak as it suggests in a real fire scenario, the MWCNT/PUF would result in a more rapidly developing fire. This is likely not true for two reasons. First, this MWCNT/PUF Cone data (Figure 18 and Table 2) was not normalized, as suggested by Zammarano et al [53], since the surface area for MWCNT/PUF was not quantitatively measured and changed significantly throughout the test. Based on qualitative observations, the HRR for the first peak could be reduced by a factor of two while the second peak may only be slightly reduced. The result is the first peak for MWCNT/PUF would have a maximum HRR and time to peak similar to PUF and the second peak would then become the PHRR, which has a time to peak similar to PUF. Secondly, the Cone data does not capture the real impact of the PUF pool fire since there is no pool fire. Since a pool fire can approximately increase the fire threat (as calculated from HRR, THR, and burn time) of a burning product (e.g. upholstered furniture) by 35 % [59] and the MWCNT coating prevents pool formation than it is assumed that in a real fire replacing PUF with MWCNT/PUF would decrease the HRR from a product by 35 %.

Within the measurement uncertainty, the CNF and MWCNT coatings have a similar impact on the fire performance of the BF (Figure 19 and Table 3). The MWCNT coating causes a 2.8 times increase in PHRR (157 kW/m^2 ± 17 kW/m^2 as compared to 57 kW/m^2 ± 9 kW/m^2) and 3.6 times increase in THR (2.9 MJ/m^2 ± 0.3 MJ/m^2 as compared to 0.8 MJ/m^2 ± 0.1 MJ/m^2) (Figure 19 and Table 3).

3.2.3. MMT-based nanocoatings

TGA was used to determine the actual mass of MMT and coating deposited onto the substrates. The MMT coatings increased the mass of the PUF substrate by 3.2 mass fraction % ± 0.6 mass fraction % and the BF substrate by 6.9 mass fraction % ± 1.4 mass fraction %, of which, 66 mass fraction % ± 15 mass fraction % is MMT. The total CNF content relative to the substrate mass is 4.6 mass fraction % ± 1.0 mass fraction % for PUF and 2.1 mass fraction % ± 0.2 mass fraction % for PUF, which is a typical loading level of MMT incorporated into polymers (not a coating) to improve the polymer's fire performance. Unlike these other nanocomposites, the CNFs in these coatings are concentrated at the surface rather than randomly dispersed and distributed throughout the polymer matrix [31].

The HRR data indicated that the MMT coatings improved the fire performance of foam (Figure 18 and Table 2). The HRR curves for MMT/PUF and PUF consist of two peaks. The time to maximum HRR each peak was 12 % ± 3 % earlier for PUF (21 s ± 2 s and 75 s ± 3 s as compared to 34 s ± 2 s and 78 s ± 2 s for MWCNT/PUF). The maximum measured HRR during the test (PHRR) was 35 % ± 3 % lower for MWCNT/PUF (620 kW/m^2 ± 26 kW/m^2 as compared to 403 kW/m^2 ± 10 kW/m^2). The THR and total burn time was also lower for MWCNT/PUF (21 % ± 3 % and 25 % ± 3 %, respectively).

Similar to PUF, MMT/PUF collpased to form a pool fire. However, the collapse was slower than PUF, but fast enough that the shape was retained for less time than the CNF/PUF and MWCNT/PUF. The result of the slower collapse is a first peak that was more similar to what was observed for CNF/PUF. The higher HRR minimum value between the first and second HRR peak, a significantly higher HRR maximum for the second peak (compared to MWCNT/PUF and CNF/PUF), and a second peak with attributes more similar to PUF are a reflection of the MMT providing little protection in the form of a char and the MMT/PUF is burning similar to a normal PUF pool fire. The inability of the MMT to form a good protective char stems from the pool fire being too shallow to allow char formation [56]. Unlike, CNF/PUF and MWCNT/PUF it is unlikely in a real fire scenario the MMT/PUF would have the potentially 35% further decrease in HRR because MMT/PUF collapses to form a pool fire [59].

Within measurement uncertainty, the CNF, MWCNT, and MMT coatings have a very similar impact on the fire performance of the BF (Figure 19 and Table 3). The MMT coating causes a 2.7 times increase in PHRR (155 kJ/m^2 ± 16 kJ/m^2 as compared to 57 kJ/m^2 ± 9 kJ/m^2) and 3.7 times increase in THR (155 MJ/m^2 ± 16 MJ/m^2 as compared to 0.8 MJ/m^2 ± 0.1 MJ/m^2) (Figure 19 and Table 3).

Table 2. Cone Calorimetry data of the washed uncoated and coated PUF organized from highest to lowest in PHRR (**PHRR in bold**). Values reported with 2σ uncertainty.

	Peak 1 HRR (kW/m^2)	Peak 1 Time (s)	Peak 2 HRR (kW/m^2)	Peak 2 Time (s)	THR (MJ/m^2)	Residue Mass Fraction %	Burn time (s)
PUF	224 ± 12	32 ± 2	**620 ± 26**	75 ± 3	33 ± 2	2.2 ± 0.1	140 ± 2
CNF/PUF	**371 ± 10**	27 ± 2	348 ± 10	83 ± 2	26 ± 1	11.0 ± 0.4	110 ± 2
MWCNT/PUF	**391 ± 10**	34 ± 2	403 ± 10	78 ± 2	26 ± 1	11.1 ± 0.4	105 ± 2
MMT/PUF	396 ± 15	36 ± 2	**515 ± 15**	86 ± 2	31 ± 2	2.6 ± 0.3	111 ± 2

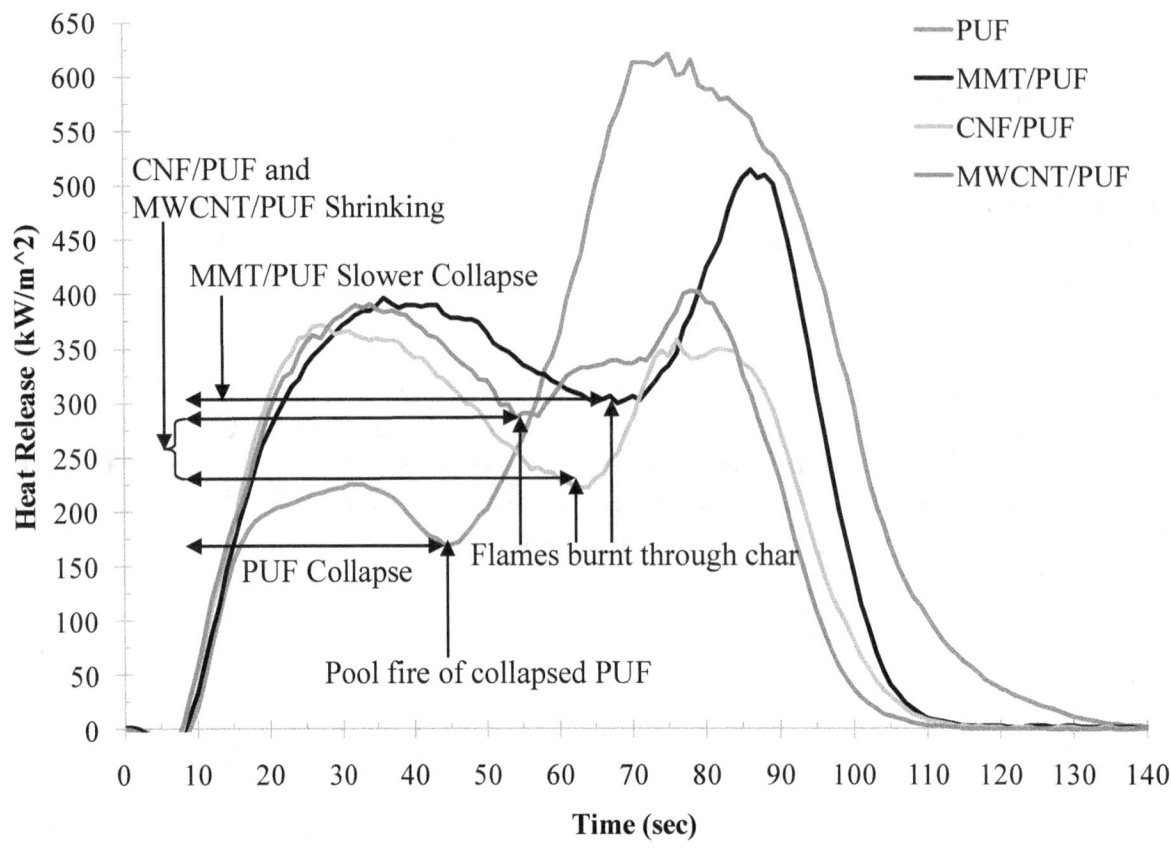

Figure 18. HRR curves indicate the MWCNT and CNF coatings significantly reduce PHRR, THR, and total burn time, but MMT coatings increase flammability, as compared to pure PUF. The 2σ uncertainty is ± 5% in HRR and ± 2 s in time.

Table 3. Cone Calorimetry data of the uncoated and coated BF organized from highest to lowest in PHRR. Values reported with 2σ uncertainty.

	Peak 1		THR (MJ/m^2)	Burn time (s)
	HRR (kW/m^2)	Time (s)		
MWCNT/BF	157 ± 17	19 ± 2	2.9 ± 0.3	65 ± 2
MMT/BF	155 ± 16	15 ± 2	2.7 ± 0.4	60 ± 2
CNF/BF	149 ± 15	16 ± 2	3.1 ± 0.3	65 ± 2
BF	57 ± 9	25 ± 3	0.8 ± 0.1	50 ± 2

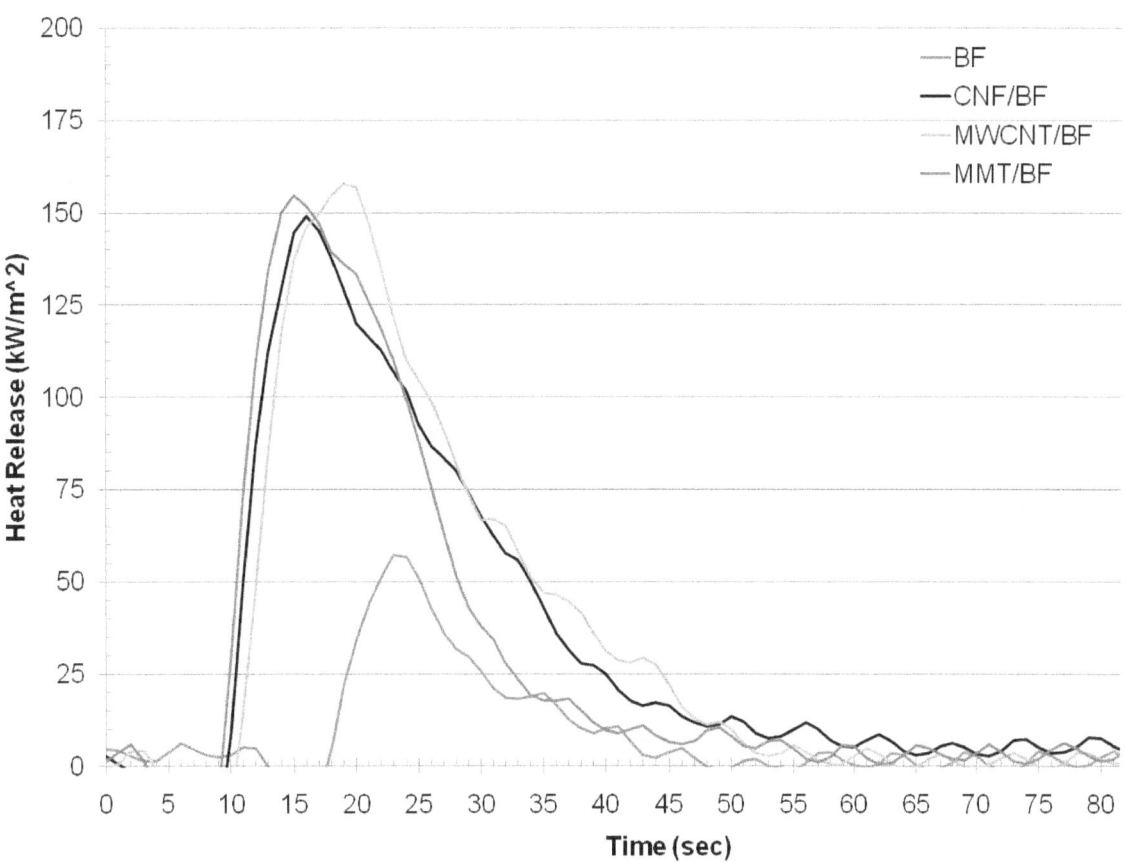

Figure 19. HRR curves of the uncoated and coated BF. All nanoparticle coatings deteriorate the fire performance of the BF. The 2σ uncertainty is ± 5% in HRR and ± 2 s in time.

4.4. Comparison to other flame retarding technologies

A previous study reported a 35 % reduction (Table 4) in PHRR for PUF using 4 mass fraction % CNFs embedded in the PUF (CNFs were added to the foam recipe) [29]. In comparison, the LbL fabricated CNF/PUF specimen has a 20 % greater reduction in PHRR using 57 mass fraction % less CNFs. In other words, incorporating CNF as a coating rather than directly into the polyurethane foam yields a three times greater reduction in PUF flammability. This information may be of particular interest to foam manufacturers, as it is assumed the post-manufacturing coating of CNFs and using less CNFs will be easier and more cost effective to implement than incorporating CNFs into the foam recipe.

Najafi-Mohajeri used the Cone to measure the impact of 17 flame retardant additive packages (five non-halogen, four halogen, and seven halogen-phosphorous) on the flammability of a standard PUF [57]. These additive packages are commercially available and reported to be commonly used by the PUF industry. The five non-halogens reduced the PUF PHRR and THR by an average of 15 % and 14 %, respectively, at a 3.3 mass fraction % averaged loading (Table 4). The four halogens reduced the PUF PHRR and THR by an average of 31 % and 16 %, respectively, at a 20 mass fraction % averaged loading. The seven halogen-phosphorous systems reduced the PUF PHRR and THR by an average of 14 % and 7 %, respectively, at a 28 mass fraction % averaged loading.

Also using a Cone, Price [58] measured the flammability impact of incorporating melamine-based flame retardants into PUF (Table 4). The exact composition of the materials is unknown as they were purchased from a supplier. The melamine and melamine chlorate phosphate blend reduced the PHRR by 10 % and 15 %, respectively. As an alternative to using flame retardants, the authors also measured the impact of using fire blocking barrier fabrics to reduce the PUF flammability. Of the six specimens tested, the best performing combination was wrapping the standard PUF with zirconium hexafluoride flame retardant treated wool (FR-wool), which gave a 29 % reduction in PHRR. This FR-wool also gave the greatest reduction in PHRR of the flame retardant foams (32 %), which was quite similar to what was reported for wrapping the standard PUF with this FR-wool. These results suggest the fire performance benefits gained by using these flame retardants are partially mitigated by the FR-wool.

Compared to these competitor flame retardant systems and the experimental CNF embedded in foam system, the nanoparticle LbL coatings developed in this project delivered a 14 % to 65 % larger reduction in PHRR and THR using 46 % to 95 % less FR [57,58]. To normalize the % PHRR reduction for the differences in FR content, the % PHRR values were divided by the FR loading level (Table 4). For example, a value of 0.5 % was calculated for the non-halogen FRs by dividing 15 % (% PHRR reduction relative to PUF) by 4 % (the FR loading level). This mass normalized % PHRR is the percent PHRR reduction measured for 1 mass fraction % loading of the FR. The FRs that yielded the lowest impact were the commercial non-halogen and halogen-phosphorous FRs with a 0.5 % reduction in PHRR for 1 mass fraction % loading of these FRs. The commercial halogens FRs yielded a slightly better reduction with a value of 1.6 %. The best impact from an embedded FR was the experimental CNF system that yielded a 5 times to 17 times larger reduction compared to the 17 commercial FRs, which is comparable to the impact of the worst LbL coating FR (MMT, 7.6 % reduction in % PHRR for 1 mass fraction % MMT). The greatest impact was measured for the CNF LbL coatings with a 17 times to 50 times greater % PHRR reduction (25 % reduction in % PHRR for 1 mass fraction % CNF) and the MWCNT LbL coatings with a 13 times to 39 times greater % PHRR reduction (19 % reduction in % PHRR for 1 mass fraction % MWCNT), relative to the commercial FRs. The only FR used in both the LbL coating and embedded was the CNF and the % PHRR reduction was 3 times when the CNF were used in the coating. The take message from this data is these nanoparticles yield a greater improvement in the fire resistance of polyurethane foam than the commonly used commercial FRs and using the nanoparticles in the LbL coating will give a greater improvement in the foam fire resistance than the embedded approach. Therefore, the LbL nanoparticle technologies deserve serious consideration for developing cost-effective fire safe polyurethane foam.

Table 4. Reduction in Cone data (relative to pure PUF) caused by FR in a LbL coating on and embedded in foam. Adjusting for different FR loadings, the LbL approach can result in the largest reduction in PHRR. No uncertainty was reported for the literature values. Experimental values reported with 2σ uncertainty.

Flame Retardant Location	FR (mass fraction % loading)	% Reduction relative to PUF		Mass FR normalized % PHRR reduction
		PHRR	THR	
In coating on foam	CNF (1.6% ± 0.1%)	(40 ± 3)%	(21 ± 3)%	25 %
	MWCNT (1.7% ± 0.1%)	(35 ± 3)%	(21 ± 2)%	19 %
	MMT (2.1% ± 0.2%)	(16 ± 3)%	(6 ± 1)%	7.6 %
Embedded in foam	CNF (4%)	34 %		8.5 %
	4 Halogen FRs (20%)	31 %	16 %	1.6 %
	5 Non-halogen FRs (4%)	15 %	14 %	0.5 %
	8 Halogen-Phosphorous FRs (28%)	14 %	7 %	0.5 %

3.5. Nanoparticle release from stressing

3.5.1. Measurements methodology

CNFs and MWCNTs released from stressing were quantified using UV-VIS. Collection and preparation of the nanoparticle containing suspensions for analysis had a significant impact on the accuracy and repeatability of the measurement. The initial methods developed for these measurements was based on analyzing the CNF containing suspensions as-received from the stressing experiments. After sonicating the simulated chewing/CNF suspension (0.050 mass fraction % ± 0.001 mass fraction % CNF and 0.9 mass fraction % ± 0.08 mass fraction % sodium chloride in DI) for 4 h, the UV-VIS absorbance value (using a 1 mm ES quartz cuvette) was significantly dependent on sonication time and the time between sonication and analysis. This dependence after long sonication times suggested the CNFs were not forming a well dispersed and stable suspension. To facilitate better CNF dispersion and distribution, more stable suspension, and decrease in sonication time, SDS (of 0.0075 mass fraction % ± 0.02 mass fraction %) was added (prior to sonication) to this simulated chewing/CNF suspension. Incorporation of SDS resulted in a stable absorbance value after sonicating for 2 h at 90 % of maximum amplitude. The sonication time was not impacted by further increasing the SDS concentration to 0.23 mass fraction % ± 0.02 mass fraction %. In contrast, without sodium chloride in the suspension (simulated wear and tear), the higher SDS concentration yielded an equally stable absorbance value in only 1 h of sonciation. Since the only difference in the suspensions is sodium chloride, it was assumed the instability in the simulated chewing suspensions was a result of salt induced CNF aggregation.

Three suspensions were prepared and analyzed to demonstrate the impact of sodium chloride on the absorbance value for CNF and MWCNT (267 nm) and the mitigation using dialysis (Figure 20). The absorbance value of a CNF suspension prepared without sodium chloride (0.050 mass fraction % ± 0.001 mass fraction % CNF, 0.23 mass fraction % ± 0.02 mass fraction % SDS, and 100 mL DI) is 2.18 ± 0.10 (Sodium chloride free spectrum). Adding sodium chloride (0.9 mass fraction % ± 0.08 mass fraction %) decreased the absorbance by 23% (1.66 ± 0.10, Before dialysis spectrum). Removing sodium chloride through 24 h of dialysis, reduced the conductivity to a value similar to DI (16,000 µS ± 300 µS reduced to < 0.5 µS) and resulted in an absorbance value similar to the sodium chloride free samples (2.16 ± 0.10, After dialysis spectrum). All spectra were collected after 1 h of sonication at 90 % maximum amplitude using a 10 mm ES quartz cuvette, except for the "Before Dialysis" which was sonicated for 4 h to obtain a stable absorbance value for this suspension. SDS was added immediately before sonication.

By adding SDS and removing sodium chloride (simulated chewing extract only), repeatable and steady state UV-VIS absorbance values were obtained after 1 h of sonication. The CNF or MWCNT concentration in the stressed suspensions was based on a Beer's Law calibration curve (CNF curve provided in Figure 21) constructed from the measured absorbance, at the maximum absorbance band (267 nm), of five prepared nanoparticle calibration suspensions. The absorbance values and the calibration curves constructed from a sodium chloride free and sodium chloride removed by dialysis suspensions were identical, which indicates that dialysis is an effective nondestructive method to remove sodium chloride. The lower limit detection of nanoparticles using this methodology is 0.00010 mass fraction % ± 0.00004 mass fraction %.

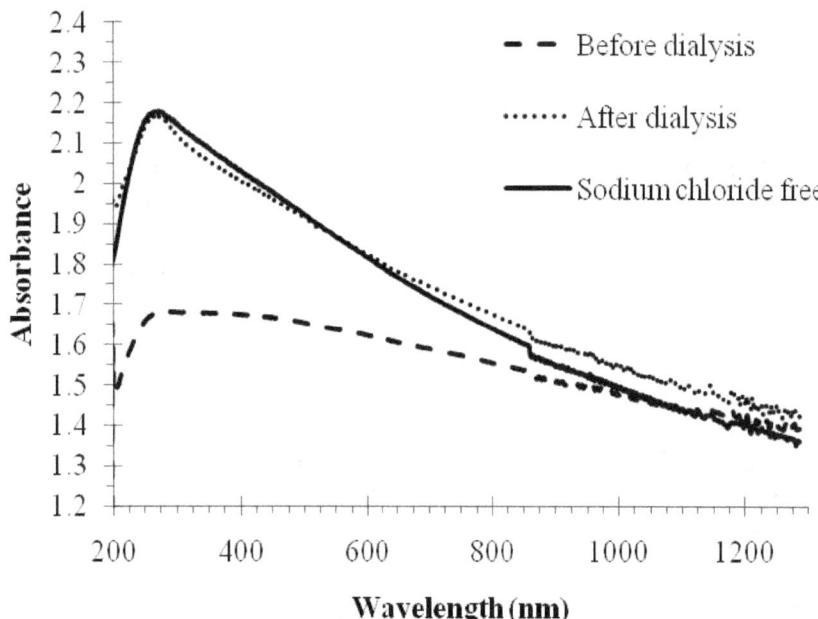

Figure 20. UV-VIS absorbance plot of a CNF/DI/SDS suspension prepared without sodium chloride, with sodium chloride, and the sodium chloride removed by dialysis. Dialysis was critical to obtaining accurate and repeatable absorbance values at the MWCNT and CNF peak maximum (267 nm) for the simulated chewing suspensions.

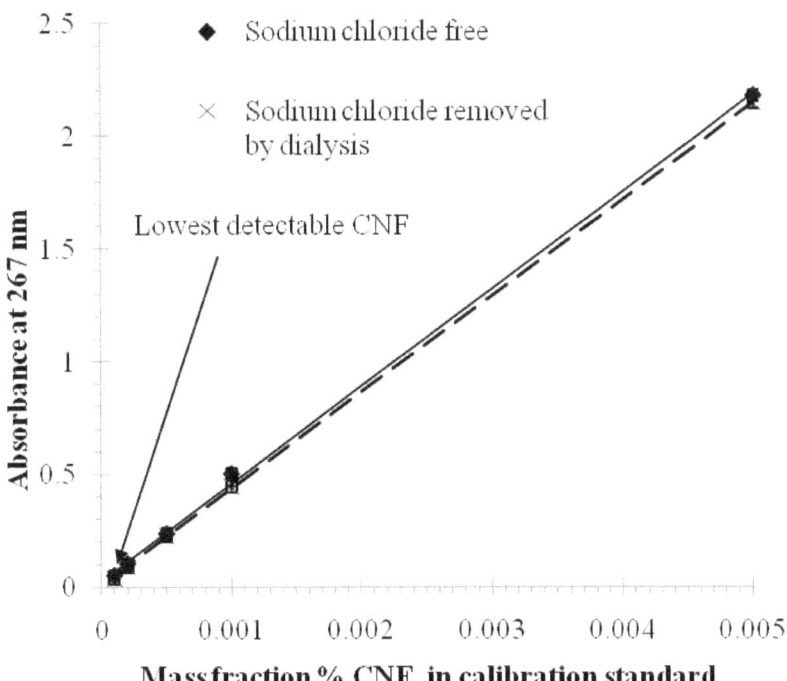

Figure 21. A Beer's Law curve was constructed from the UV-VIS absorbance value of several CNF or MWCNT calibration standards (CNF provided here). CNF and MWCNT concentration was quantified based on the measured UV-VIS absorbance value of a stressed suspension and a Beer's Law curve. The 2σ uncertainty is ± 5 % of the CNF or MWCNT mass fraction value.

Clay released during stressing was quantified using a Beer's Law calibration curves constructed from the ICP-OES intensity of three mass fraction % MMT calibration standards in diluted nitric acid and DI. The amount of Al, Fe, Mg, and Na (elements) in MMT was determined using a standard addition method. More specifically, the ICP-OES intensity of five diluted MMT/DI solutions with different known amounts of each element was measured and plotted as a function of element type and concentration (Figure 22). The element concentration at zero intensity is the amount of that element in MMT in a diluted nitric acid solution containing 0.0010 mass fraction % ± 0.0005 mass fraction % MMT (Figure 23). The Beer's Law calibration curves were constructed from the ICP-OES intensity of the diluted nitric acid solution containing 0.0010 mass fraction % ± 0.0005 mass fraction % MMT and three lower mass fraction % MMT solutions that were created by dilution of this initial solution with more diluted nitric acid (Figure 24). In contrast to the UV-VIS analysis, the stressed solutions were analyzed as-received by ICP-OES because the sodium chloride did not impact the analysis. However, the Na element is not used to quantify MMT concentration since the sodium chloride also contributes to the Na element ICP-OES intensity. Mg element was also not used to quantify MMT concentration since Mg can be found in PUF and BF. Therefore, MMT concentration is an average based on Fe and Al ICP-OES intensity.

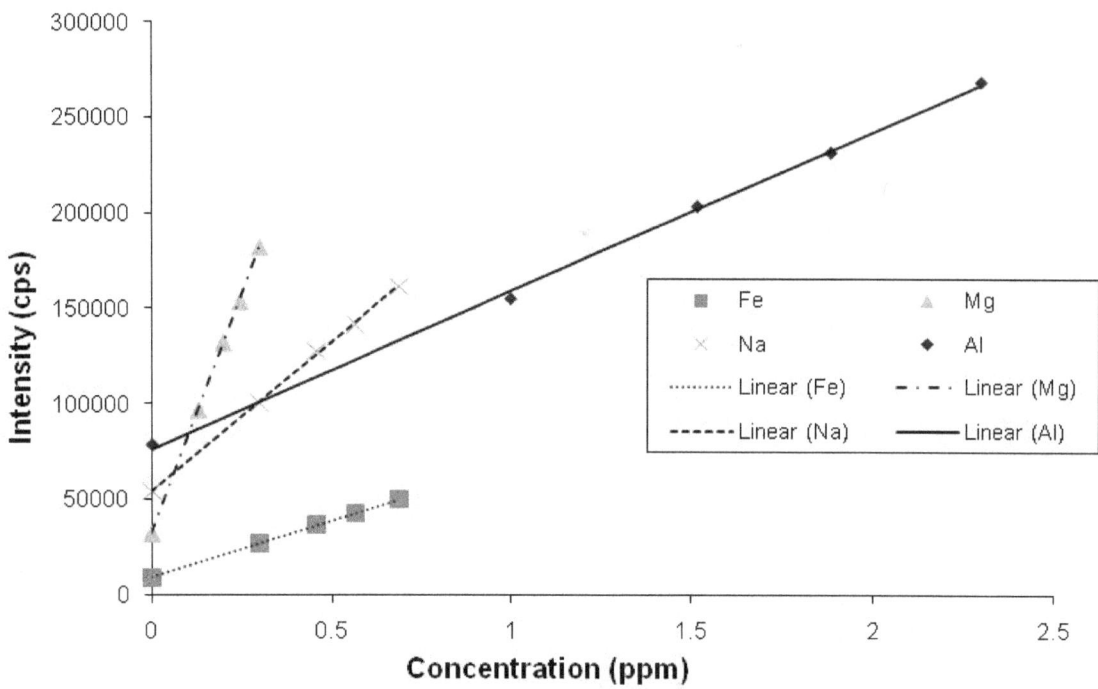

Figure 22. ICP-OES intensity as a function of element type and concentration using a diluted nitric acid/DI solution containing 0.0010 mass fraction % ± 0.0005 mass fraction % MMT and five different concentrations of each element. The concentration at an intensity of 0 cps is the amount of each element in MMT. The 2σ uncertainty is 5 % of the value (error bars are smaller than the data markers).

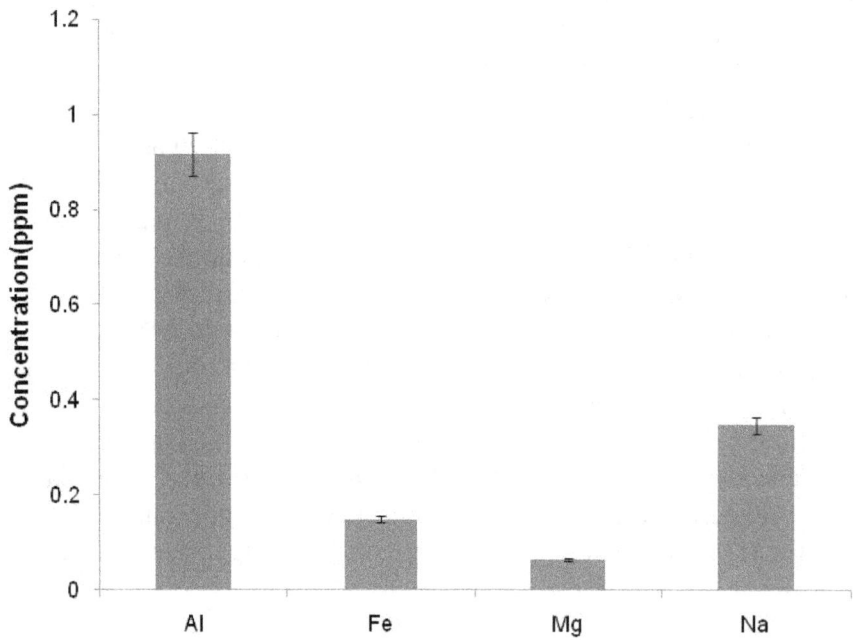

Figure 23. The amount of each element in MMT in a diluted nitric acid/DI solution containing 0.0010 mass fraction % ± 0.0005 mass fraction % MMT in diluted. The error bars represent a 2σ standard uncertainty in the concentration.

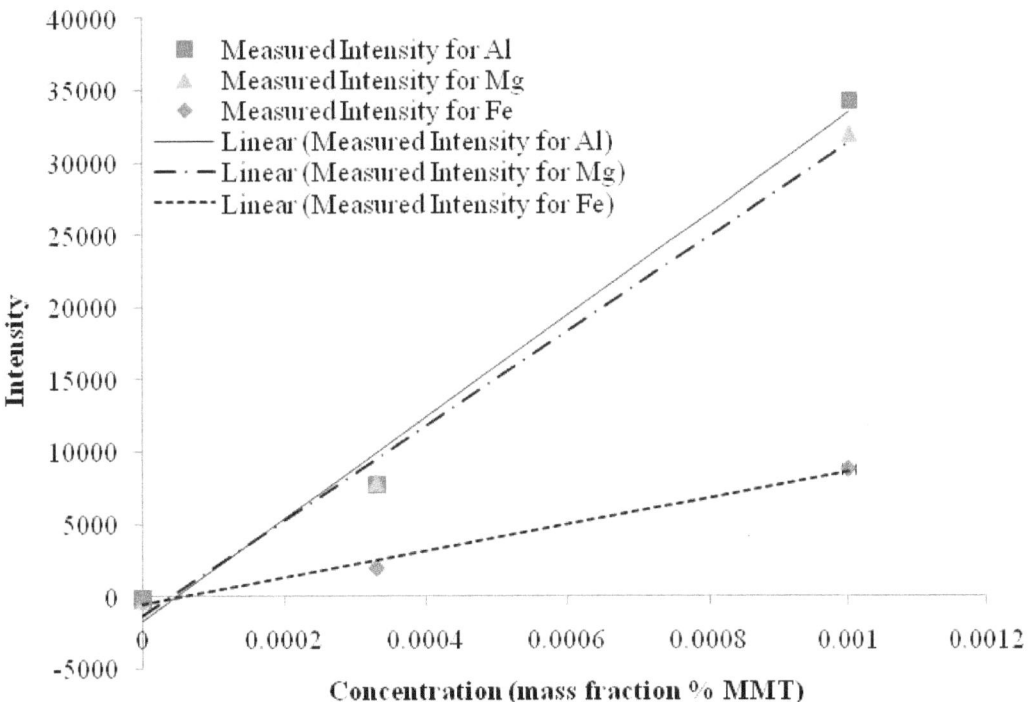

Figure 24. Beer's Law calibration curve of ICP-OES intensity as a function of MMT concentration. The 2σ uncertainty is 5 % of the value (error bars are smaller than the data markers).

3.5.2. Stress induced nanoparticle release

The entire coated substrate (in a sealed plastic bag) was used for the simulated wear and tear experiments. The released nanoparticles are washed out of the bag using a SDS/DI solution (described in the experimental section). The same wash solution was used for 10 stressed samples so that the wash solution contained the nanoparticles released from 10 samples. This approach was used to increase the concentration of nanoparticles in the analyzed suspensions as the initial measurements indicated the nanoparticles released from a single substrate was below the detection limit of the instruments. Experiments conducted with known concentrations of nanoparticles indicated that within the detection limit of the measurement tools, all of the released nanoparticles were recovered using this methodology.

The nanoparticle release values from the simulated wear and tear experiments are provided in Table 5. There were two general trends measured in the wear and tear studies. One was a higher release using CNFs and MWCNTs and a lower release using MMT. The other was more nanoparticles were released from the BF than the PUF substrate. More specifically, the release of CNFs and MWCNTs was 3.5 times higher from the BF than the PUF, which was an order of magnitude higher than MMTs released from either substrate. The highest and lowest release values came from CNF/BF and MMT/PUF, respectively, which differ by more than two orders of magnitude.

SEM images of all the released nanoparticles indicate the nanoparticles are embedded in the polymer coating and/or adhered to the substrate. In other words, the higher release values from

BF may result from the BF being less durable than PUF to the mechanical stresses and not an indication of the nanoparticle coating durability. Presumably, the higher amount of CNFs released, as compared to the MMTs and MWCNTs, may result from the CNFs not being completely embedded in the polymer matrix (Figure 8 to Figure 11) and therefore easier to break away when the substrate is stressed. It is assumed that MMTs and MWCNTs were completely embedded in the coating because they are smaller and the TL is a more effective coating approach. Reducing the release of CNFs using a TL approach is currently under investigation.

A small portion of the coated substrates were used for the simulated chewing experiments (4.6 mass fraction % ± 0.1 mass fraction % of the entire coated substrate). The extraction solutions from 40 specimens were combined, sodium chloride was removed by dialysis and then the suspensions were stabilized by SDS. Similar to the simulated wear and tear experiments, this methodology yielded the most accurate and repeatable measurements, and combining the release from several substrates increased the nanoparticle content in the suspension to values above the LDL of the tool.

The nanoparticle release values from the simulated chewing experiments are provided in Table 6. There were three general trends observed from the simulated chewing experiments. One was the highest release came when using MWCNTs and similar lower release values came from using CNFs and MMTs. Another was there appeared to be a bias, all be it smaller than in the wear and tear experiments, toward higher release from the BF. Lastly, compared to wear and tear, the release values were lower and closer in the chewing experiments.

For simulated chewing, the highest release values (MWCNTs) were an order of magnitude greater than the lowest release values (CNFs or MMTs on PUF) (Table 6). The MWCNTs release values were independent of the substrate and 3 times higher than the release of CNFs and MMTs from the BF. However, the CNFs and MMTs release values did depend on the substrate with a 3 times higher release from the BF than the PUF. Within the uncertainty of the measurements, the release of CNFs and MMTs were similar for a given substrate. The lowest chewing release values were comparable to the highest wear and tear release values. Therefore, as a general rule of thumb, the release from chewing was an order of magnitude greater than from wear and tear.

Similar to wear and tear, the nanoparticles released from the simulated chewing were all embedded in the polymer matrix. The reason for higher CNF and MMT release from the BF may be a reflection of substrate durability. It is assumed that the simulated saliva causes the non-woven fabric to swell and separate. This combined with agitation may be sufficient stress to cause fragments of materials to wash out of the fabric. This "swelling" may also be the root cause tor the higher release values for the chewing experiments. All the materials used in the coating can be dissolved in water and therefore if given sufficient time it is reasonable to expect that the coating in water will expand and potentially wash off the substrate. Reducing the release of nanoparticle by crossslinking the coating is currently under investigation. The higher release of MWCNT may be a result of its size and shape. The long fiber length of CNFs and the large flat surface of MMTs may provide sufficient interaction with the surrounding nanoparticles to help hold them in the coating.

Table 5. Simulated wear and tear nanoparticles released relative to the total nanoparticle content on the specimen (organized from highest to lowest in nanoparticle release). Values reported with 2σ uncertainty.

Substrate	Mass per specimen (mg)	Mass fraction % of total nanoparticle content
CNF/BF	0.037 ± 0.004	**0.040 ± 0.004**
MWCNT/BF	0.025 ± 0.003	**0.027 ± 0.003**
CNF/PUF	0.021 ± 0.002	**0.010 ± 0.001**
MWCNT/PUF	0.019 ± 0.002	**0.0085 ± 0.0008**
MMT/BF	0.0018 ± 0.0002	**0.00095 ± 0.00009**
MMT/PUF	0.0011 ± 0.0001	**0.00028 ± 0.00002**

Table 6. Simulated chewing nanoparticles released relative to the total nanoparticle content on the specimen (organized from highest to lowest in nanoparticle release). Values reported with 2σ uncertainty.

Substrate	Mass per specimen (mg)	Mass fraction % of total nanoparticle
MWCNT/PUF	0.047 ± 0.004	**0.46 ± 0.04**
MWCNT/BF	0.046 ± 0.005	**0.42 ± 0.04**
CNF/BF	0.014 ± 0.001	**0.14 ± 0.01**
MMT/BF	0.037 ± 0.004	**0.13 ± 0.01**
CNF/PUF	0.0052 ± 0.0005	**0.053 ± 0.005**
MMT/PUF	0.0054 ± 0.0005	**0.043 ± 0.004**

4. Conclusion

The characteristics of the CNF-based and MWCNT-based coatings on PUF were distinctly different with the MMT-based coatings having features similar to both of these nanocoatings. More specifically, the MWCNT coatings completely and uniformly covered the entire surface of the PUF and, other than a few sparsely distributed tens of micron sized MWCNT aggregates, the coating appeared smooth and featureless at lower magnifications. In contrast, the CNF-based coatings were rougher with an appearance more similar to a fibrous network rather than a smooth, uniform coating. These coatings covered all the PUF surfaces. Due to its large dimensions the CNFs tended to deposit as groups rather than individual fibers, which resulted in regions of high CNF concentration (tens of microns in size) and regions of no CNFs (less than a few microns in size). The highly aggregated regions contained fibers that were only partially embedded in the polymer coating. The regions of smaller aggregates (islands) generally contained fibers almost completely embedded in the polymer coating. The MMT coatings appeared rough due to regions of high clay aggregation (similar to CNF). These clay aggregates likely formed in the depositing solution and therefore deposited as large aggregates. The clay appeared to be completely embedded in the polymer matrix, which, in the non-aggregate regions, resulted in a coating that appeared smooth (similar to MWCNT). The CNF and MWCNT loading in the coatings were the same (50 mass fraction % ± 1 mass fraction %), but the CNF coating was 28% ± 8% thinner than the MWCNT coating (359 nm ± 36 nm as compared to 440 nm ± 47 nm, respectively). In contrast, the MMT coatings were much thicker with a high variability of thickness (1000 nm ± 450 nm) and contained a higher concentration of nanoparticles (66 mass fraction% ± 16 mass fraction %).

One of the major factors impacting the physical characteristics of the coatings is the stability of the coating preparation suspensions. Within 4 h of stopping sonication, the CNFs begin to settle to the bottom of the depositing solution as indicated by a color gradient (darker at the bottom) in the CNF suspension. During deposition the CNF suspension appears very uniform. However, it is proposed that the reason the CNFs deposit as aggregates is because the CNFs had already begun to aggregate in the suspension. In comparison, the MWCNT suspensions remain visibly uniform over two days. This difference in quality of the suspensions is most likely a result of the smaller dimensions and/or the PEI functionalizing of the MWCNT. For the MMT, the suspension appeared to be fairly stable with no clay settling for several days. It is assumed the clay surface aggregation was more a result of poor clay dispersion in the DI than of poor suspension stability.

Even though the physical characteristics of the coatings were quite different, the improvement in PUF fire performance due to the CNF and MWCNT LbL fabricated coatings were similar and 10 times to 50 times better (on a mass basis) than 17 commercial FRs commonly used in PUF. The MMT LbL fabricated coatings had less of an impact on fire performance (relative to the MWCNT and CNF); however, they were still 3 times to 17 times better than the commercial FRs, which is similar to what has been reported for CNFs embedded in the foam. This data suggests that these nanoparticles could potentially result in significantly less flammable PUF, but an at least 3 times greater improvement could be achieved if these nanoparticles were added through a LbL fabricated coating rather than embedded in the foam. This post-manufacturing process route to increasing fire resistance performance of foam is attractive because it potentially has no impact of the foam manufacturing process.

The primary reason for the drastic decrease in flammability is that the coatings prevent foam melt dripping and increase char formation. In a real fire scenario, the formation of a pool fire, which is created by melting PUF, but not CNF/PUF or MWCNT/PUF, approximately increases the fire threat (as calculated from HRR, THR, and burn time) of the burning product by 35% [59]. This impact is not captured in Cone data because there is no product (soft furnishing, etc.) for the pool fire to pose an additional flux upon. Therefore, the Cone data is a conservative measure of the improved fire performance created by the MWCNT and CNF coating; however, the actual benefit in real fires from using a CNF/PUF or MWCNT/PUF instead of PUF, could be 35% greater than reported here. MMT coating also reduces the PHHR and THR, but not as significantly as CNF and MWCNT due to the inability of the MMT to maintain the shape and prevent the pool fire. For all nanoparticles, the nanoparticle coated BFs were more flammable than the pure BF. This may be due to the non-woven BF falling apart during the coating process.

The release of nanoparticles was in general an order of magnitude higher from simulated chewing than simulated wear and tear. These coatings (constructed of water dissolvable and suspendable materials) may be sensitive to moist stresses (e.g., washing and cleaning). To reduce nanoparticle release in simulated chewing, additional durability experiments should be conducted (e.g., crosslink the coating). Release was highest from the BF presumably due to the lower durability of the non-woven fabric design, as compared to the thermoset foam. A more durable barrier fabric should be considered for further investigation of this LbL technology (e.g., woven fabric). The simulated wear and tear release values were CNFs ~ MWCNTs (BF) > CNFs ~ MWCNTs (PUF) >> MMT (BF and PUF). The simulated chewing release values were MWCNT (BF and PUF) > CNF ~ MMT (BF) > CNF ~ MMT (PUF). To create a more durable coating (e.g., less release), the coating could be crosslinked or a TL approach could be used. It is important to keep in mind that a more durable coating may not be necessary from a risk exposure point of view since there has not been a toxicity evaluation of the materials released from these substrates.

5. Future Research

This research has laid the foundation for using LbL to fabricate coatings on foam and barrier fabrics using a range of nanoparticles and other performance enhancing additives. We are currently fabricating and analyzing cellulosic fiber coatings and mixed additive coatings on both foam and barrier fabrics.

6. References

[1] Hall JR. Total cost of fire in the United States. National Fire Protection Association report; March 2010. Summary available from:
http://www.nfpa.org/assets/files/PDF/totalcostsum.pdf.

[2] Hall JR, Harwood B. The national estimates approach to U.S. fire statistics. Fire Technology. 1989;25(2):99-113.

[3] Miller D, Chowdhury R, Greene M. 2005-2007 Residential Fire Loss Estimates. Consumer Product Safety Commission report; August 2010. Available from:
http://www.cpsc.gov/library/fire07.pdf.

[4] Ahrens M. Home fires that began with upholstered furniture. National Fire Protection Association report; May 2008. Summary available from:
http://www.nfpa.org/assets/files/PDF/UpholsteredExecutiveSum.pdf.

[5] 16 CFR 1632 Standard for the flammability of mattresses and mattress pads. Consumer Product Safety Commission; May 1991. Available from:
http://www.cpsc.gov/businfo/testmatt.pdf.

[6] 16 CFR 1633 Standard for the flammability (open flame) of mattress sets. Consumer Product Safety Commission regulation; March 2007. Available from:
http://www.cpsc.gov/businfo/frnotices/fr06/mattsets.pdf.

[7] 16 CFR Part 1634, Standard for the flammability of residential upholstered furniture (proposed rule). Consumer Product Safety Commission; March 2008. Available from:
http://www.cpsc.gov/businfo/frnotices/fr08/furnflamm.pdf.

[8] Gann RG, Stackler KD, Ruitberg S, Guthrie WF, Levenson MS. Technical Note 1436 Relative Ignition Propensity of Test Market Cigarettes. National Institute of Technology report: Available from: http://www.bfrl.nist.gov/pdf/final_report.pdf.

[9] http://echa.europa.eu/reach_en.asp

[10] http://ec.europa.eu/environment/ecolabel/

[11] Decher G. Chapter 1 Polyelectrolyte multilayers: An overview. Mutlilayer Thin Films: Sequential Assembly of Nanocomposite materials: G Dechner and JB Schlenoff (Eds.); Wiley-VCH: Weinheim, Germany, 2003.

[12] Podsiadlo P, Shim BS, Kotov, NA. Polymer/clay and polymer/carbon nanotube hybrid organic-inorganic multilayered composites made by sequential layering of nanometer scale films. Coordination Chemical Reviews. 2009;253:2835-2851.

[13] Li YC, Schulz J, Mannen S, Delhom C, Condon R, Chang S, Zammarano M, Grunlan JC. Flame retardant behavior of polyelectrolyte-clay thin film assemblies on cotton fabric. ACS Nano. 2010;4(6):3325-3337.

[14] Li F, Peng ZH, Zhang LL, Yao LS, Xuan L. Photoalignemnt of liquid crystals in a hydrogen bonding directed Layer-by-Layer ultrathin films. Liquid Crystals. 2009;36:43-51.

[15] Bergbreiter DE, Chance BS. "Click"-Based covalent Layer-by-Layer assembly on polyethylene using water soluble polymeric reagents. Macromolecules. 2007;40:5337-5343.

[16] Mermut O, Barrett CJ. Effects of charge density and counterions on the assembly of polyelectrolyte multilayers. Journal of Physical Chemistry B. 2003;1072525-2530.

[17] Chang L, Kong X, Wang F, Wang L, Shen J. Layer-by-Layer assembly of poly(N-acryloyl-N'-propylpiperazine) and poly(acrylic acid): effect of pH and temperature. Thin Solid Films. 2008;516:2125-2129.

[18] Priolo MA, Gamboa D, Grunlan JC. Transparent clay-polymer nano brick wall assemblies with tailorable oxygen barrier. ACS Applied Materials & Interfaces. 2010;2:312-320.

[19] Aoki P, Volpati D, Riul A, Caetano W, Constantino CJL. Layer-by-layer technique as a new approach to produce nanostructured films containing phospholipids as transducers in sensing applications. Langmuir. 2009;25:2331-2338.

[20] Fu JH, Ji J, Fan DZ, SHen JC. Construction of Antibacterial multilayer films containg nanosilver via Layer-by-Layer assembly of heparin and chitosan-silver ions complex. Journal Biomedical Material Research Part A. 2008;79A:665-674.

[21] Hiller J, Mendelsohn JD, Rubner MF. Reversibly erasable nanoporous antireflection coatings from polyelectrolyte multilayers. Nature Materials. 2002;1:59-63.

[22] Podasiadlo P, Michel M, Lee J, Verplogen E, Kam NWS, Ball V, Lee J, Qi Y, Hart AJ, Hammond PT, Kotov NA. Exponential growth of LBL films with incorporated inorganic sheets. Nano Letters. 2008;8:1762-1770.

[23] Li YC, Grunlan JC. Polyelectrolyet/nanosilicate thin-film assemblies: Influence of pH on growth, mechanical behavior, and flammability. ACS Applied Materials & Interfaces. 2009;1:2338-2347.

[24] Davis RD, Gilman JW, Vanderhart DL. Processing degradation of polyamide 6/montmorillonite clay nanocomposites and clay organic modifier. Polymer Degradation and Stability. 2003;79:111-121.

[25] Gilman JW, Bourbigot S, Shields JR, Nyden M, Kashiwagi T, Davis RD, Vanderhart DL, Demory W, Wilkie CA, Morgan AB, Harris J, Lyon RE. High throughput methods for polymer nanocomposites research: extrusion, NMR characterization, and flammability property screening. Journal of Materials Sceince. 2003;38:4451-4460.

[26] Davis RD, Lyon RE, Takemori MT, Eidelman N. Chapter 16 High throughput techniques for fire resistant materials development, Fire Retardancy of Polymeric Materials 2nd Edition. CA Wilkie and AB Morgan (Eds.): Taylor and Francis, Boca Raton, FL. 2010; p 421-451.

[27] Jiang DD. Polymer Nanocomposites, Fire Retardancy of Polymeric Materials 2nd Edition. CA Wilkie and AB Morgan (Eds.): Taylor and Francis, Boca Raton, FL. 2010; p 261-299.

[28] Gilman JW, Davis RD, Shields JR, and Harris RH. High Throughput Flammability Characterization Using Gradient Heat Flux Fields. J ASTM Int. 2005;2(9):1-11.

[29] Zammarano M, Kramer RH, Harris RH, Ohlemiller TJ, Shields JR, Rahatekar SS, Lacerda S, Gilman JW. Flammability reduction of flexible polyurethane foams via carbon network formation. Polymers for Advanced Materials. 2008;19:588-595.

[30] Tibbets GG, Max LL, Strong KL, Rice BP. A review of the fabrication and properties of vapor-grown nanofiber/polymer composites. Composites Science and Technology. 2007;64:1709-1718.

[31] Kashiwagi T, Du F, Douglas JF, Winey KI, Harris RH, Shields JR. Nanoparticle networks reduce the flammability of polymer nanocomposites. Nature Materials. 2005;4:928-933.

[32] Nikolaev P, Bronikowski MJ, Bradley RK, Rohmund F, Colbert DT, Smith KA, Smalley RE. Gas-phase catalytic growth of single-walled carbon nanotubes from carbon monoxide. Chemical Physics Letters, 1999;313(1-2):91-97.

[33] Coleman JN, Khan U, Blau WJ, Gun'ko YK. Small but strong: A review of the mechanical properties of carbon nanotube-polymer composites. Carbon, 2006;44(9):1624-1652.

[34] Yu M.F, Files BS, Arepalli S, Ruoff RS. Tensile loading of ropes of single wall carbon nanotubes and their mechanical properties. Physical Review Letters. 2000;84(24):5552-5555.

[35] Thess A, Lee R, Nikolaev P, Dai HJ, Petit P, Robert J, Xu CH, Lee, YH, Kim SG, Rinzler AG, Colbert DT, Scuseria GE, Tomanek D, Fischer JE, Smalley RE. Crystalline ropes of metallic carbon nanotubes. Science. 1996;273(5274):483-487.

[36] Yu C, Shi L, Yao Z, Li DY, Majumdar A. Thermal conductance and thermopower of an individual single-wall carbon nanotube. Nano Letters. 2005;5(9):1842-1846.

[37] Fantini C, Cassimiro J, Peressinotto VST, Plentz F, Souza AG, Furtado CA, Santos AP. Investigation of the light emission efficiency of single-wall carbon nanotubes wrapped with different surfactants. Chemical Physics Letters. 2009;473(1-3):96-101.

[38] Grossiord N, Loos J, Regev O, Koning CE. Toolbox for dispersing carbon nanotubes into polymers to get conductive nanocomposites. Chemistry of Materials. 2006;18(5):1089-1099.

[39] Moore VC, Strano MS, Haroz EH, Hauge RH, Smalley RE, Schmidt J, Talmon Y. Individually suspended single-walled carbon nanotubes in various surfactants. Nano Letters. 2003;3(10):1379-1382.

[40] Yu C, Kim YS, Kim D, Grunlan JC. Thermoelectric Behavior of Segregated-Network Polymer Nanocomposites. Nano Letters. 2008;8(12):4428-4432.

[41] Kang YK, Lee OS, Deria P, Kim SH, Park TH, Bonnell DA, Saven JG, Therien MJ. Helical Wrapping of Single-Walled Carbon Nanotubes by Water Soluble Poly(p-phenyleneethynylene). Nano Letters. 2009;9(4):1414-1418.

[42] Zou JH, Liu LW, Chen H, Khondaker SI, McCullough RD, Huo Q, Zhai L. Dispersion of pristine carbon nanotubes using conjugated block copolymers. Advanced Materials. 2008;20(11):2055-2071.

[43] Liu L, Grunlan JC. Clay assisted dispersion of carbon nanotubes in conductive epoxy nanocomposites. Advanced Functional Materials. 2007;17(14):2343-2348.

[44] Zhu J, Yudasaka M, Zhang MF, Iijima S. Dispersing carbon nanotubes in water: A noncovalent and nonorganic way. Journal of Physical Chemistry B. 2004;108(31):11317-11320.

[45] Liu L, Etika KC, Liao K, Hess L, Bergbreiter DE, Grunlan JC. Comparison of Covalently and Noncovalently Functionalized Carbon Nanotubes in Epoxy. Macromol Rapid Communications. 2009;30(8):627-632.

[46] Lafuente E, Callejas MA, Sainz R, Benito AM, Maser WK, Sanjuan ML, Saurel D, de Teresa JM, Martinez MT. The influence of single-walled carbon nanotube functionalization on the electronic properties of their polyaniline composites. Carbon. 2008;46(14):1909-1917.

[47] Bartholome C, Miaudet P, Derre A, Maugey M, Roubeau O, Zakri C, Poulin P. Influence of surface functionalization on the thermal and electrical properties of nanotube-PVA composites. Composites Science and Technology. 2008;68(12):2568-2573.

[48] Certain commercial equipment, instruments or materials are identified in this paper in order to specify the experimental procedure adequately. Such identification is not intended to imply recommendation or endorsement by the National Institute of Standards and Technology, nor is it intended to imply that the materials or equipment identified are necessarily the best available for this purpose.

[49] The policy of NIST is to use metric units of measurement in all its publications, and to provide statements of uncertainty for all original measurements. In this document however, data from organizations outside NIST are shown, which may include measurements in non-metric units or measurements without uncertainty statements.

[50] In this document, we have provided link(s) to website(s) that may have information of interest to our users. NIST does not necessarily endorse the views expressed or the facts presented on these sites. Further, NIST does not endorse any commercial products that may be advertised or available on these sites.

[51] Future Foam Inc, 2451 Cypress Way, Fullerton CA 92831-5103, 714-871-5103.

[52] Liao K, Wan A, Batteas JD, Bergbreiter DE. Superhydrophobic Surfaces Formed Using Layer-by-layer Self-Assembly with Aminated Multi-walled Carbon Nanotubes. Langmuir. 2008;24(8):4245-4253.

[53] M. Zammarano. BFRL Foam Flammability Consortia homepage, http://www.bfrl.nist.gov/866/foam/. [cited 2010 July 20].

[54] Drum X Model 85 Tissue Culture Rotator, Lab Instruments PO Box 1835, Rockville, MD 20850.

[55] ES1 Type 1, ES Quartz, 1 mm LP cuvette, NSG Precision Cells, Farmingdale, NY 11735, www.nsgpci.com.

[56] Kashiwagi T, Gilman JW, Butler KH, Harris RH, Shields RH. Flame retardant mechanism of silica gel/silica. Fire and Materials. 2000;24:277-283.

[57] N. Najafi-Mohajeri, C. Jayakody, G.L. Nelson. Cone Calorimetric Analysis of Modified Polyurethane Elastomers and Foams with Flame-Retardant Additives. Fire and Polymers. American Chemical Society; 2010; p 79-89. Available from: http://dx.doi.org/10.1021/bk-2001-0797.ch007

[58] D. Price, Y. Liu, R. Hull, G.J. Milnes, B.A. Kandola, A.R. Horrocks. Burning behavior of fabric/polyurethane foam combinations in the cone calorimeter. Polym. Intern. 49 (2000) 1153-1157.

[59] Pitts WM, Haspias G, Macatangga P. Fire spread and growth on flexible polyurethane foam. Eastern States Section of the Combustion Institute Fall 2009. University of Maryland, College Park, MD. October 18-21, 2009; p. 11-22.

www.ingramcontent.com/pod-product-compliance
Lightning Source LLC
Chambersburg PA
CBHW081740170526
45167CB00009B/3886